ClearRevise®

OCR Cambridge Nationals
IT J836 (R050)

Illustrated revision and practice

Published by
PG Online Limited
The Old Coach House
35 Main Road
Tolpuddle
Dorset
DT2 7EW
United Kingdom

sales@pgonline.co.uk
www.clearrevise.com
www.pgonline.co.uk
2024

PREFACE

Absolute clarity! That's the aim.

This is everything you need to ace the examined component in this course and beam with pride. Each topic is laid out in a beautifully illustrated format that is clear, approachable and as concise and simple as possible.

Each section of the IT specification is clearly indicated to help you cross-reference your revision. The checklist on the contents pages will help you keep track of what you have already worked through and what's left before the big day.

We have included worked exam-style questions with answers for every topic. This helps you understand where marks are coming from and to see the theory at work for yourself in an exam situation. There is also a set of exam-style questions at the end of each section for you to practise writing answers for. You can check your answers against those given at the end of the book.

LEVELS OF LEARNING

Based on the degree to which you are able to truly understand a new topic, we recommend that you work in stages. Start by reading a short explanation of something, then try and recall what you've just read. This has limited effect if you stop there but it aids the next stage. Question everything. Write down your own summary and then complete and mark a related exam-style question. Cover up the answers if necessary but learn from them once you've seen them. Lastly, teach someone else. Explain the topic in a way that they can understand. Have a go at the different practice questions – they offer an insight into how and where marks are awarded.

Design and artwork: Mike Bloys / PG Online Ltd

First edition 2024   10 9 8 7 6 5 4 3 2 1
A catalogue entry for this book is available from the British Library
ISBN: 978-1-916518-20-9
Copyright © PG Online 2024
All rights reserved
No part of this publication may be reproduced, stored in a retrieval system, or transmitted in any form or by any means without the prior written permission of the copyright owner.

This product is made of material from well-managed FSC® certified forests and from recycled materials.

Printed by Bell & Bain Ltd, Glasgow, UK.

THE SCIENCE OF REVISION

Illustrations and words

Research has shown that revising with words and pictures doubles the quality of responses by students.[1] This is known as 'dual-coding' because it provides two ways of fetching the information from our brain. The improvement in responses is particularly apparent in students when they are asked to apply their knowledge to different problems. Recall, application and judgement are all specifically and carefully assessed in public examination questions.

Retrieval of information

Retrieval practice encourages students to come up with answers to questions.[2] The closer the question is to one you might see in a real examination, the better. Also, the closer the environment in which a student revises is to the 'examination environment', the better. Students who had a test 2–7 days away did 30% better using retrieval practice than students who simply read, or repeatedly reread material. Students who were expected to teach the content to someone else after their revision period did better still.[3] What was found to be most interesting in other studies is that students using retrieval methods and testing for revision were also more resilient to the introduction of stress.[4]

Ebbinghaus' forgetting curve and spaced learning

Ebbinghaus' 140-year-old study examined the rate at which we forget things over time. The findings still hold true. However, the act of forgetting facts and techniques and relearning them is what cements them into the brain.[5] Spacing out revision is more effective than cramming – we know that, but students should also know that the space between revisiting material should vary depending on how far away the examination is. A cyclical approach is required. An examination 12 months away necessitates revisiting covered material about once a month. A test in 30 days should have topics revisited every 3 days – intervals of roughly a tenth of the time available.[6]

Summary

Students: the more tests and past questions you do, in an environment as close to examination conditions as possible, the better you are likely to perform on the day. If you prefer to listen to music while you revise, tunes without lyrics will be far less detrimental to your memory and retention. Silence is most effective.[5] If you choose to study with friends, choose carefully – effort is contagious.[7]

---

1. Mayer, R. E., & Anderson, R. B. (1991). Animations need narrations: An experimental test of dual-coding hypothesis. *Journal of Education Psychology*, (83)4, 484–490.
2. Roediger III, H. L., & Karpicke, J.D. (2006). Test-enhanced learning: Taking memory tests improves long-term retention. *Psychological Science*, 17(3), 249–255.
3. Nestojko, J., Bui, D., Kornell, N. & Bjork, E. (2014). Expecting to teach enhances learning and organisation of knowledge in free recall of text passages. *Memory and Cognition*, 42(7), 1038–1048.
4. Smith, A. M., Floerke, V. A., & Thomas, A. K. (2016) Retrieval practice protects memory against acute stress. *Science*, 354(6315), 1046–1048.
5. Perham, N., & Currie, H. (2014). Does listening to preferred music improve comprehension performance? *Applied Cognitive Psychology*, 28(2), 279–284.
6. Cepeda, N. J., Vul, E., Rohrer, D., Wixted, J. T. & Pashler, H. (2008). Spacing effects in learning a temporal ridgeline of optimal retention. *Psychological Science*, 19(11), 1095–1102.
7. Busch, B. & Watson, E. (2019), *The Science of Learning*, 1st ed. Routledge.

CONTENTS

R050: IT in the digital world
Topic Area 1: Design tools

Specification point ☑
1.1	Flow charts	2 ☐
1.1	Mind maps	4 ☐
1.1	Visualisation diagrams	7 ☐
1.1	Wireframes	8 ☐
	Examination practice: Topic 1	10 ☐

Topic Area 2: Human Computer Interface (HCI) in everyday life

Specification point ☑
2.1	The purpose, importance and use of HCI in application areas	12 ☐
2.2	Hardware considerations	16 ☐
2.2	Software considerations	18 ☐
2.3	HCI use in Digital platforms	20 ☐
2.4	User interaction methods	21 ☐
	Examination practice: Topic 2	22 ☐

Topic Area 3: Data and testing

Specification point ☑
3.1 - 3.2.1	Information and data	23 ☐
3.2 - 3.2.4	Validation and Verification	24 ☐
3.2.3	Data validation tools	25 ☐
3.3	Data collection methods	26 ☐
3.4	Storage of collected data	28 ☐
3.5	Testing	30 ☐
	Examination practice: Topic 3	32 ☐

Topic Area 4: Cyber-security and legislation

Specification point

4	Cyber-security and legislation	33
4.1	Threats	34
4.2	Impacts of cyber-security attacks	36
4.3	Prevention measures	37
4.4	Legislation related to IT systems	40
4.4	Health and Safety at Work Act	42
	Examination practice: Topic 4	43

Topic Area 5: Digital communications

Specification point

5.1	Types of digital communications	44
5.2	Software	46
5.3	Digital devices	48
5.4.1	Types of distribution channel	49
5.4.2	Distribution channel connectivity	51
5.5	Audience demographics	53
	Examination practice: Topic 5	54

Topic Area 6: Internet of Everything (IoE)

Specification point

6.1	Internet of Everything (IoE)	56
6.1	Interactivity between the four pillars of IoE	58
6.2	Applications of IoE in everyday life	60
	Examination practice Topic 6	66

Examination practice answers	67
Levels-based mark schemes for extended response questions	76
Index	78
Examination tips	**81**

MARK ALLOCATIONS

Green mark allocations[1] on answers to in-text questions throughout this guide help to indicate where marks are gained within the answers. A bracketed '1' e.g. [1] = one valid point worthy of a mark. In longer answer questions, a mark is given based on the whole response. In these answers, a tick mark [✓] indicates that a valid point has been made. For a mark, a judgement should be made using the levels-based mark scheme on **page 76**. There are often many more points to make than there are marks available so you have more opportunity to max out your answers than you may think.

ACKNOWLEDGMENTS

Every effort has been made to trace and acknowledge ownership of copyright. The publishers will be happy to make any future amendments with copyright owners that it has not been possible to contact. The publisher would like to thank the following companies and individuals who granted permission for the use of their images in this textbook.

All Sections
Images: © Shutterstock
iPad image © Alexey Boldin / Shutterstock.com
SmartWatch image © A. Aleksandravicius / Shutterstock.com
Kindle image © A. Aleksandravicius / Shutterstock.com
Western Digital hard drive image © DAMRONG RATTANAPONG / Shutterstock.com
Samsung Galaxy Flip phone image © AronX / Shutterstock.com
The Bluetooth® word mark and logos are registered trademarks owned by Bluetooth SIG,Inc.
Bluetooth symbol image © bamby st / Shutterstock.com

TOPICS FOR EXAM
Unit R050 IT in the digital world

Information about the externally assessed exam

Written exam: 1 hour and 30 minutes
70 marks
Section A: 15 marks
Section B: 55 marks

Specification coverage
Knowledge of IT in the digital world, topic areas 1-6.

Topic Area 1: Design tools

Topic Area 2: Human Computer Interface (HCI) in everyday life

Topic Area 3: Data and testing

Topic Area 4: Cyber-security and legislation

Topic Area 5: Digital communications

Topic Area 6: Internet of Everything (IoE)

Questions
Section A: A range of closed response, multiple choice and short answer questions.

Section B: Scenario based questions which require knowledge and understanding from all the topic areas in R050. A hand-drawn/sketch question worth 8 marks, such as mind maps, flow charts or visualisation diagrams, will be given. An extended response question worth 9 marks will also be given.

1.1

FLOW CHARTS

A **flow chart** uses symbols and connecting lines to show the steps in a **process**.

Flow chart uses

There are many uses for flow charts. For example, they could show a process that happens when a form is submitted on a website.

You need to know four design tools:
- Flow charts
- Mind maps
- Visualisation diagrams
- Wireframes

Flow chart components

Flow charts use common symbols so that they are easily understood. The symbols are connected via flow lines. This shows the direction of flow through the flow chart.

Symbol	Meaning
Start / End	A **start/end symbol** is used at the **start** of the flow chart and the **end** of the flow chart. A flow chart should only have one start and one end symbol. The end symbol is also called a **terminator**.
Process	A **process box** shows that some processing will happen such as a calculation being performed.
Decision	A **decision symbol** is used when a choice of direction needs to be taken. For example, "Is the switch on?". Decision boxes usually have two outputs – **True** or **False**. However, sometimes they can have more than this.
Input / Output	**Input/Output symbols** show where a user enters data or the results of processing are displayed. *Input examples*: PRESS BUTTON, INPUT age *Output examples*: OUTPUT age, DISPLAY RESULTS.
Flow lines → TRUE / YES → FALSE / NO →	**Flow lines** show the direction of travel through a flow chart. The arrow on the end shows the direction. Flow lines are usually given a label of **Yes**/**No** or **True**/**False** when they come out of a decision symbol. Some decision boxes have more than two arrows that come out from them - for example, a different direction for each of the days of the week.

Creating flow charts

Flow charts can represent any process. This flow chart shows the process for turning a security light on or off. It uses two decision symbols and two input/output symbols.

Flow charts are easy to draw by hand. You can also draw them in software programs.

Some software programs are specifically designed to draw flow charts. These include **Microsoft Visio** or **draw.io**. **Word processing** and **presentation software** usually allows flow charts to be created by inserting relevant symbols and arrows.

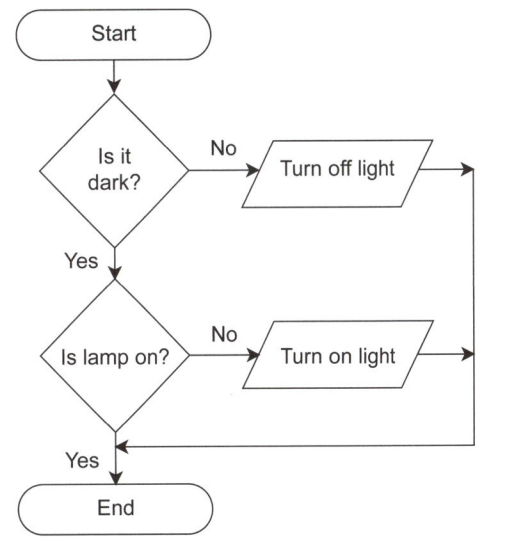

Advantages and disadvantages of flow charts

➕ Advantages	➖ Disadvantages
• Flow charts are quick to draw by hand. • Digital flow charts are quick and easy to create and edit. With good software, flow lines will automatically update. • Processes and steps are clearly shown.	• Specialist software may be needed to create digital flow charts. • Hand drawn flow charts are hard to change if a mistake is made. • Can be complex and difficult to read if a process is very complex. • Specific symbols need to be known to create or understand a flow chart.

1. Flow charts use arrows to show the direction of flow.
   State **three** other components that may be used in a flow chart. [3]
2. The flow chart shown at the top of this page is used to turn a security light on and off. It only checks whether the light needs to be turned on or off once.
   Describe how the flow chart could be altered so that it constantly checks whether the light needs to be turned on or off. [2]
3. One disadvantage of using flow charts is that they become complex and difficult to read for large problems. State **two** other disadvantages of using flow charts. [1]

   1. Decision,[1] Input,[1] Output,[1] Process,[1] Start,[1] End.[1]
   2. The end symbol needs to be removed[1] and the Yes arrow (from the 'is lamp on?' decision symbol) will go into the 'Is it dark' decision symbol[1] to form a loop.[1]
   3. Specialist software may be required to create one.[1] Hand drawn flow charts are hard to change if a mistake is noticed.[1] Specific symbols need to be understood to create or understand a flow chart.[1]

1.1

MIND MAPS

Mind maps show information that has been organised so that similar ideas or concepts are grouped together with links showing the relationships between the items.

Mind map components

The components used in a mind maps are:
- A **central idea** or **theme**
- **Nodes**
- **Sub-nodes**
- **Branches**

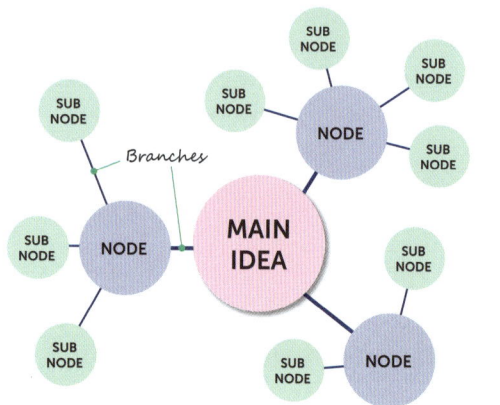

Related ideas on a mind map may be given the same colour to help them stand out. Images may also be used to aid understanding.

A central node is used for the main idea.

Sub-nodes are nodes that branch off other nodes.

Creating mind maps

Mind maps are easily created by hand with just a pen and paper. General software such as presentation software or a word-processor often have features to draw the circles and lines needed to create mind maps. However, using this type of software is often time consuming. **Mind mapping software** such as FreeMind helps to develop mind maps quickly.

Advantages and disadvantages of mind maps

➕ Advantages
- They are very quick to draw by hand.
- They help in generating lots of ideas.
- They are easy for others to understand.

➖ Disadvantages
- If there are too many nodes they become confusing.
- They give key points, but not key details and information.
- They are more time consuming to create than a list.

Library mind maps

Library mind maps are used for organising information. They help to show linked ideas.

Library maps start with a central node which is the focus of the mindmap. For example it could be the name of a new product or a new website.

Ideas linked to the main idea are then linked from the central node by drawing lines.

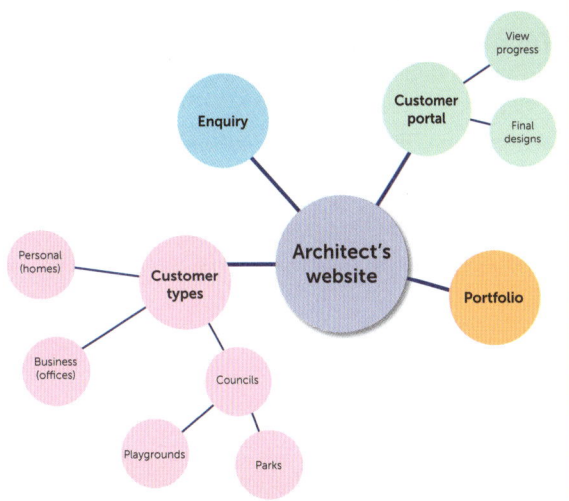

Presentation mind maps

Presentation mind maps are used to present an idea or concept to a specific audience. The layout and structure will be presented in a way that makes it easy for people to understand. The use of **colours** or even **images** may help in presentation mind maps.

The example shows a presentation mind map made for people to see the different forms of transport they could use in a city. It could be used in a slide in a business presentation or on a page of a tourist leaflet.

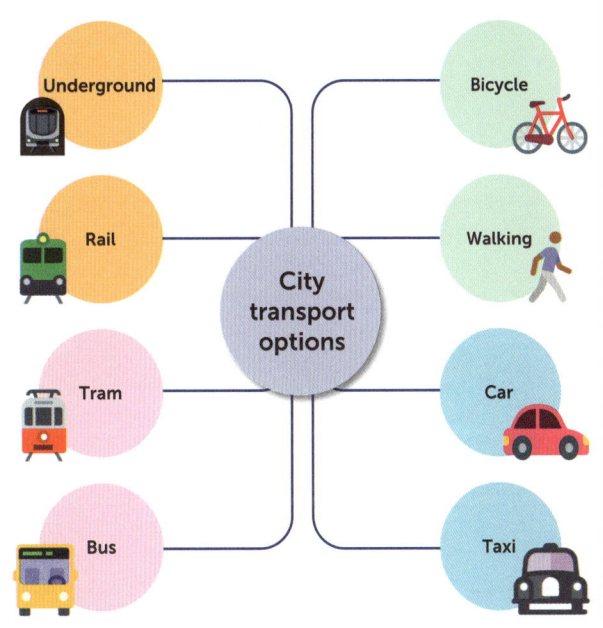

1. State **one** disadvantage of using a mind map. [1]
2. An airline is training their staff. All staff need to be aware of key areas of training that they are expected to know. Explain a suitable type of mind map that could be used. [2]

   1. They can become confusing if there are lots of nodes/sub-nodes.[1] Key details/information are likely to be missing.[1] They are more time consuming to create than a list.[1]

   2. A presentation mind map[1] would be suitable as they are made with the audience in mind[1] and will make use of design/layout features which help understanding.[1]

1.1

Tunnel timeline

Tunnel timeline mind maps show the steps required to solve a problem or create a project strategy. Tunnel timeline mind maps show how a goal will be achieved. The problem is placed in the central node. The steps to solve the problem or achieve the goal are put in the sub-nodes.

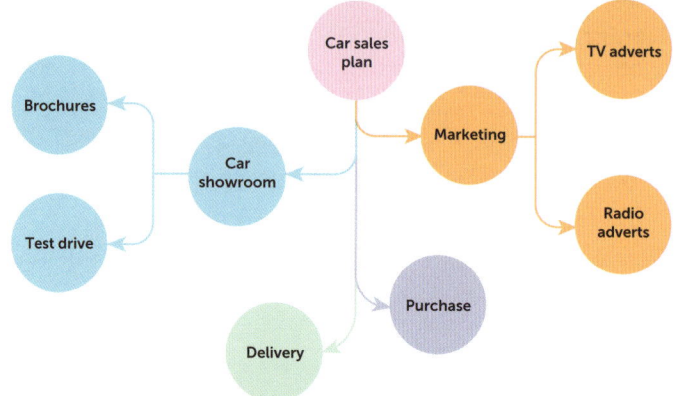

3. A tunnel timeline is one type of mind map. Give **two** other types of mind maps. [2]
4. Create a tunnel mind map for a travel agent organising a day trip to a theme park. [3]

3. *Library mind map,*[1] *presentation mind map.*[1]

4.

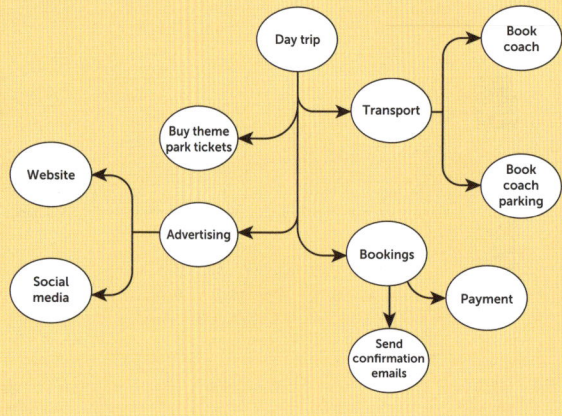

Sub nodes all come from first node.[1]
Sub nodes show the steps involved.[1]
Nodes/sub nodes are relevant to a travel agent organising a day trip.[1]

VISUALISATION DIAGRAMS

Visualisation diagrams show a sketch or diagram of the look and layout of a finished product such as a web page or poster. A visualisation diagram may use simple boxes for different elements, or contain a more refined sketch or design created on a computer. They may be created in black and white or colour.

Components of a visualisation diagram

The components of a visualisation diagram usually include:

- Size and location of images, text and buttons
- Sketches of key images
- Text for headings and boxes to indicate other text, images or buttons
- Annotation describing colours, font styles and size

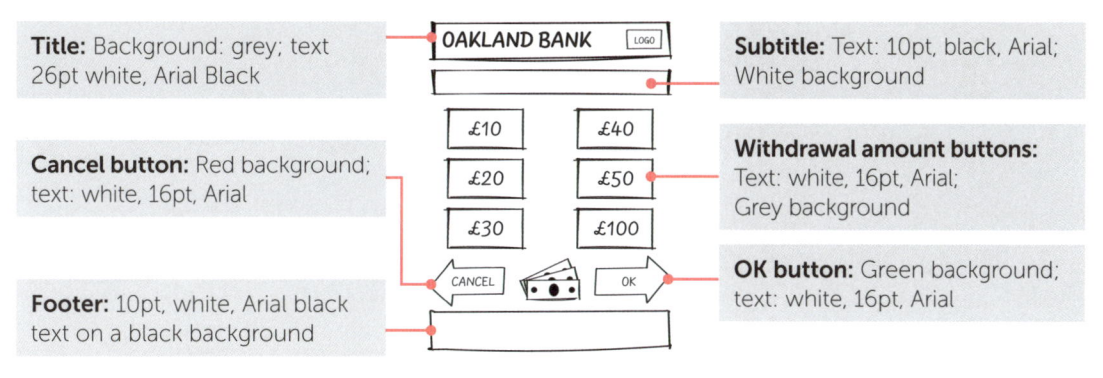

Title: Background: grey; text 26pt white, Arial Black

Cancel button: Red background; text: white, 16pt, Arial

Footer: 10pt, white, Arial black text on a black background

Subtitle: Text: 10pt, black, Arial; White background

Withdrawal amount buttons: Text: white, 16pt, Arial; Grey background

OK button: Green background; text: white, 16pt, Arial

Creating visualisation diagrams

Sketched visualisation diagrams use pens, pencils and paper. **Generic software** such as Microsoft PowerPoint or more specialist graphics or desktop publishing (DTP) software could also be used to create diagrams.

Advantages and disadvantages of visualisation diagrams

➕ Advantages	➖ Disadvantages
- Gives a good look and feel to how a product will appear when finished. - Shows layout. - Includes annotations. - Shows key details such as specific colour codes, and fonts/sizes.	- Can become very detailed and take a long time to create. - Does not show functionality. - Needs more design skills than creating a wireframe (**see page 8**). - Can be harder to edit as the diagrams are more complicated.

1.1

WIREFRAMES

A **wireframe** shows a basic layout of a product. A wireframe focuses on the functionality of a product. Functionality means what the product will do.

Wireframe components

A wireframe has minimal detail and shows the structure and layout of a product. It typically doesn't use colour or images. Wireframes are commonly used when designing websites or other user interfaces. Arrows between wireframes help to show the **navigation** the user takes through the website or product.

After a wireframe is complete, a full visualisation diagram may then be created to give a better feel for how the webpage or product will appear.

Symbols used in wireframes

The following symbols are commonly used when creating wireframes

Image		Dropdown box	
Button	ADD	Checkboxes	
Text		Radio buttons	
Search bar		Progress bar	

Wireframe layouts

Wireframes are often laid out using a **horizontal navigation** at the top of the webpage or screen. Alternatively, **vertical navigations** may be used. These are usually located on the left side of the design.

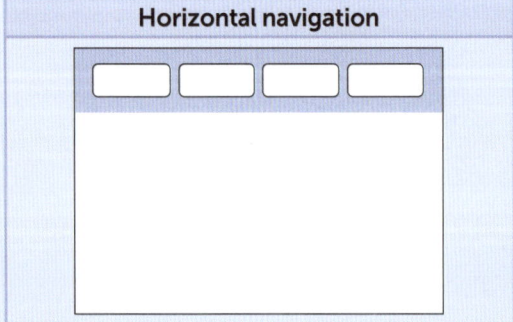
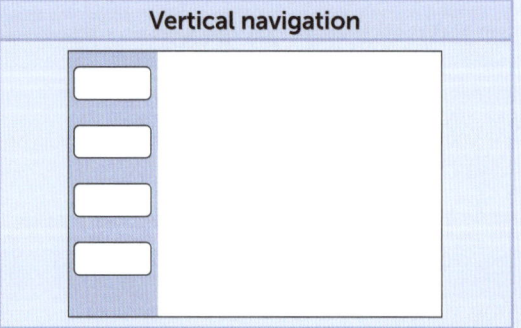

Example wireframe

The following shows a wireframe for the products page in an online shop.

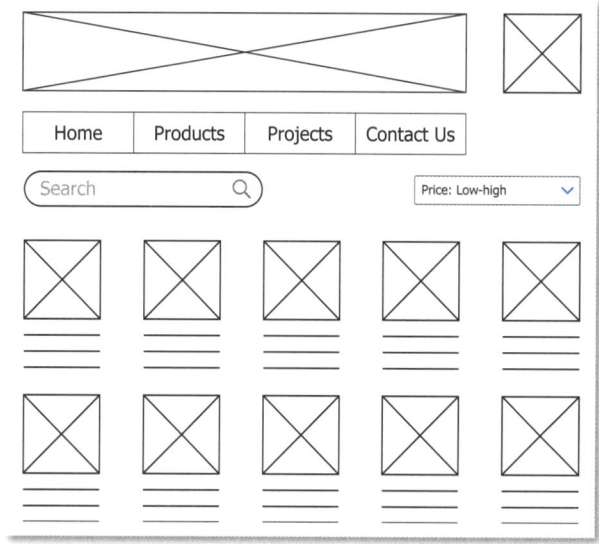

Creating wireframes

Wireframes can quickly be sketched with pen and paper.

Alternatively, generic office software such as Microsoft PowerPoint can be used. Specialist **wireframing software** will have components ready to drag and drop into designs and can also allow connections between each wireframe to be made.

Advantages and disadvantages of wireframes

➕ Advantages

- Focuses on the function of items.
- Can be drawn quickly by hand.
- Simple to understand.
- Shows key features such as layout and structure.
- Can be quick and easy to edit if drawn in software.
- Helps to show the flow through a user interface or website.
- Usually drawn in black and white.

➖ Disadvantages

- Can take longer to design using software.
- Does not show exact details such as colours or graphics.
- Does not show technical aspects of a design.

Describe **two** differences between a wireframe and a visualisation diagram. [4]

Wireframes show basic layouts[1] whereas visualisation diagrams more closely represent the final product.[1]

A wireframe is focused on functionality[1] whereas a visualisation diagram includes specific annotations[1] such as font style / font size / colours.[1] Sketches of images will be used in a visualisation diagram whereas a box with an X in it will be used in a wireframe.[1]

Topic 1

EXAMINATION PRACTICE

Section A style questions

1. Complete the table by identifying the meaning of each of the following flow chart symbols. [4]

Flow chart symbol	Meaning
parallelogram	
diamond	
rounded rectangle	
rectangle	

2. Which flow chart symbol must be present in any flow chart that is created?
   Tick (✓) **one** box only. [1]
   A – Sub-routine ☐
   B – Decision labels ☐
   C – One or more arrows indicating direction of flow ☐
   D – A flow chart title ☐

3. One type of mind map is a tunnel timeline mind map.
   Identify **one** other type of mind map [1]

4. Give **two** components used in mind maps. [2]

Section B style questions

An online retailer sells board games. They are creating a new website which will allow customers to buy the board games.

5 Before the website is built, a wireframe will be made.

  (a) Describe the purpose of a wireframe. [2]

  (b) Explain **one** disadvantage and **one** advantage of a using a wireframe rather than a visualisation diagram. [4]

The website will give information about how to play the games they sell and it will also have reviews from customers.

  (c) Create a wireframe for a product webpage that could be used to sell one board game.

  Marks will be awarded for:
  - layout;
  - content. [8]

6 When a customer is ready to complete their order, the following process occurs:
  - If the customer doesn't yet have an account they are asked to register.
  - They must enter their name, email address and password to register.
  - If the customer already has an account, they must log in with their email address and password.
  - The total cost of the order is then calculated by taking the total cost of the basket and adding a £10 delivery charge to it
  - The total cost is then shown to the user.
  - Once the customer has clicked okay then an email is sent to them.

Create a flow chart for this process.

Marks will be awarded for:
  - layout;
  - content. [8]

2.1

THE PURPOSE, IMPORTANCE AND USE OF HCI IN APPLICATION AREAS

A **Human Computer Interface** (**HCI**) allows users to interact with a computer system. HCI incorporates the devices which are used for input, such as keyboards, mice and touchscreens. An important part of HCI is the way that the **user interface** is presented and works. The **operating system** is usually responsible for this.

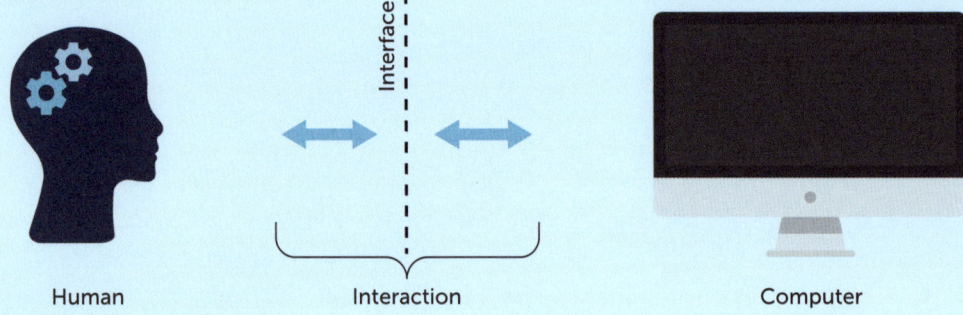

Human — Interaction — Computer

Effective use of HCI

It is important that HCIs are well designed. Imagine using a smartphone where every screen and menu was laid out differently. Alternatively, consider using a word processor where every command was listed in a single, long list. Both these examples would make it very hard for a human to interact with the computer and show poor examples of HCI.

The following areas are considered when making effective HCI for computer systems and devices:

- **Clear:** different elements of the system must be clear and easy to locate.
- **Intuitive:** the system must be logical and easy to follow. Ideally users should understand how to use the system without the use of a manual or training.
- **Consistent:** similar screens or features are always in the same place.
- **Interactive:** errors and issues are shown to the user so that they can correct them and complete the tasks they need to do.
- **Simple:** Only the items the user requires are shown on each screen.

HCI application areas

HCI is applied to a large variety of application areas.

Banking

Banking makes use of HCI with **cash machines**, **mobile banking** and **Internet banking**.

HCI usage with cash machines

- Cash machines present very few options to users.
- Colouring buttons red, yellow and green makes it more obvious which to select.
- Asking users to take their card before their cash is released prevents them forgetting their card.
- Braille dots on buttons for those with visual impairment.
- Verifying a user identity with a personal identification number (PIN) is faster for the user and only requires a numeric keypad.
- Screen layouts are consistent.
- Onscreen help makes it easier for the user to understand options.

Home appliances

Home appliances include **washing machines**, **microwave ovens** and **traditional ovens**.

HCI usage in home appliances

- Simple dials and buttons are used.
- Common symbols, such as off, play and pause, help the user understand each function.

Entertainment

Entertainment devices include **smart TVs**, **games consoles** and **smartphones**.

HCI usage in televisions

- Simple user interface that can be controlled through a remote control.
- Typically uses a menu system.
- Large icons that are easy to identify.

---

1. Give **three** features for effective use of HCI. [3]
2. Explain **two** features used on a microwave oven that demonstrate effective use of HCI. [4]

   1. Clear,[1] intuitive,[1] consistent,[1] interactive,[1] simple.[1]
   2. Large buttons[1] make them easy to press.[1] Simple and commonly used symbols/icons[1] help the user to understand the function of each item.[1] Dials are intuitive[1] for selecting cooking time.[1] A large/bright countdown timer[1] makes it easy for the user to see the remaining cooking time.[1] A beeping sound[1] helps the user know when food has been cooked.[1]

OCR Cambridge Nationals J836 (R050) **IT – Topic 2**

2.1 HCI application areas

Embedded systems

Embedded systems include **car engine control units** (**ECU**), **heating systems**, **digital watches**, **satellite navigation** (**satnav**) devices and home appliances such as **dishwashers** and **digital TVs**.

Embedded systems are usually basic digital systems that use **microcontrollers**. They usually have very simple HCIs.

HCI usage in embedded systems

- Limited options and features.
- The interface is specifically designed for one use so may make it easier for the user. For example, a lever for a car indicator or a dial to select the time on a microwave oven.
- Displays often use LEDs which are simple, bright and easy to see from a distance.

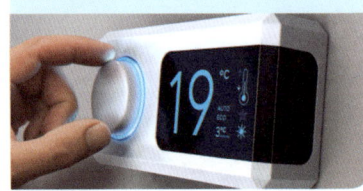

Fitness

Fitness devices range from **smart watches**, **wearable tracking devices** (such as Fitbits), **bicycle computers** and monitoring devices used on gym equipment.

HCI usage in fitness

- The devices have very specific features which helps with designing simple interfaces.
- Key information, such as distance cycled, is usually available all the time.
- Devices are easy to operate whilst exercising. Often one button press is all that is needed to change function.

Retail

Used in **Point of Sale** (**POS**) checkouts by checkout assistants. Also used with **self service checkouts** in supermarkets and fast food restaurants.

HCI usage in retail

- Simple and attractive interfaces are used for self service checkouts. These need to be usable without training.
- Large screens and buttons for customer systems.
- Systems for checkout staff have interfaces which are quick and efficient to use. They often use clear text buttons rather than graphics. Staff usually require training.

---

3. Explain **one** reason why embedded system HCIs are often simple to use. [2]
4. Explain **one** difference in the HCI used in a supermarket self-service checkout compared to the till used by a checkout assistant. [2]

3. They have one function/a limited set of features[1] which means the input hardware/method can be refined/made more intuitive.[1]

4. The self-service checkout will have a simple interface[1] that doesn't require training[1] to make it easy for customers that have never used the system before.[1] The self-service checkout will have a graphically appealing interface[1] to improve the customer experience/encourage them to shop more with the supermarket.[1] The till/POS checkout will be customised for quick/efficient use[1] by having more text/buttons and fewer graphics.[1] It may require training to use.[1]

Advantages and disadvantages of use of HCI in application areas

	➕ **Advantages**	➖ **Disadvantages**
Banking	• ATMs have simple interfaces to make transactions easy. • The interface helps to lower chances of mistakes/errors. • Displays key information clearly.	• Poorly designed HCI can have a large impact, for example money could be transferred to an incorrect bank account. • Security risks with poor design of HCI.
Embedded systems	• Very simple to use. • Used for basic functions. • Full control over the hardware allows the selection of the most appropriate input devices for each function.	• Usually limited functions. • Hard to change once manufactured, for example, extra buttons cannot be added to a car dashboard and software upgrades are typically not available.
Entertainment	• Able to customise the user interface to the user's preferences. • Engaging and attractive graphics. • Each user is able to have their own profile.	• May be inefficient, for example, searching for a film title with an on screen keyboard and a remote control. • Simple interfaces usually have limited features.
Fitness	• Very easy to navigate. • Designed to be used whilst active. • Shows key information clearly.	• Very simple functions. • Wearable devices have a limited functions due to small screen sizes.
Home appliances	• Similar advantages to embedded systems. • Screens are customised to the function of the appliance. For instance, a washing machine may have LEDs to show which program has been selected and a clear display to show time remaining.	• Similar disadvantages to embedded systems. • Limited input and output devices may make them complicated to use. For example, changing the time on an oven may require referring to an instruction manual for a sequence of key presses.
Retail	• Online access allows up to date products to appear as soon as they are available for sale. • May cut down queues in shops as customers can order at self-service checkouts. • Custom checkout designs allow checkout assistants to be more efficient.	• Must have access to networks or the Internet for the interface to work. • May result in job losses if customers are able to use self-service or shop online. • Large range of options may confuse users. • Relies on digital skills to use well. • Difficult to design well so that they can be confidently used by customers without the need for training.

5. Explain **one** consideration that a bank needs to make when using HCI in their online mobile app. [2]
6. Give **one** reason why a fitness watch's HCI needs to be easy to navigate and clear. [1]

> 5. It needs to verify entered data[1] such as the amount/account details to transfer money[1] to prevent mistakes/the money transferring to the wrong account.[1]
> It needs to be easy to use[1] as users won't have had training/read an instruction manual.[1]
>
> 6. Users will be exercising, so don't want to be distracted[1] and may need to be able to use the watch without looking at it.[1] The display needs to be clear so that the user can get key information when glancing at the watch.[1]

2.2

HARDWARE CONSIDERATIONS

When developing a device, different hardware is chosen by manufacturers. Key hardware choices include the type and size of display, the amount of memory and the type of processor that will be used.

Displays

LCD

Liquid crystal displays (**LCD**) use **liquid crystals** to change how much light passes through different areas of the screen. They may be **backlit** (such as with a television), or reflect daylight back (as used for many calculators).

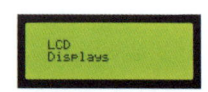

Almost all televisions, computer displays and tablets use LCD screens these days. Some use a **fluorescent backlight**. More commonly **LED backlights** are used as these are brighter.

LED

LED displays use individual **LEDs** (**Light Emitting Diodes**). They are often used in domestic appliances. The LEDs are bright allowing easier viewing than LCD displays.

LED video screens are used for outdoor screens at festivals, concerts and sports stadiums. As they are very bright, they can be viewed in daylight.

OLED

OLED (**Organic LED**) screens are thinner, more flexible and have a higher contrast to LCD screens. This makes them well suited for smartphones, especially where the screen needs to bend in a flip phone. The technology is more expensive than LCD screens.

Electronic paper

Electronic paper makes use of small pigments that raise to the top or bottom of the screen when an electric current is applied. This makes each part of the screen appear black or white. The screens are very clear in sunlight – just like real paper. Electronic paper is mostly used in e-readers. It is typically only black and white, however, some newer screens use colour.

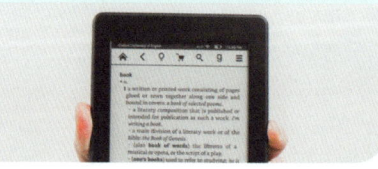

Touch screen

Touch screens allow the user to interact by touching the screen with a finger. **Capacitive touch screens** allow multiple finger touches to be sensed. This is used to detect intuitive gestures such as **pinching** to zoom in and out of a document. **Pressure sensitive touch screens** are less accurate but cheaper.

··· Displays *continued*

The **size** of a display is the **physical dimensions** of the display which is usually measured by the **diagonal size**. For example, the measurement for a 65" TV is shown. Screen sizes vary from a few centimetres for a smartwatch or mobile phone, up to many metres for advertising billboards or stadium displays.

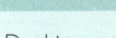

The HCI design will depend on the size of the display it is used on. Small display sizes mean there's a limited area for showing objects such as icons, buttons or windows. Small display sizes require careful thought to ensure that they can be used effectively.

> Don't confuse the **physical size** of a display with the **resolution** (such as standard definition (SD) or high definition (HD)).
>
> This unit only covers physical size.

Resources

Two key resources required for HCIs are **memory** and **processing power**. If there is insufficient processing power for the tasks of a computer or device then it will run slowly or may not work at all.

Memory

Memory, also known as **RAM** (**Random Access Memory**) is used to store apps, programs and the operating system that are currently running.

Games and graphics editing on a desktop computer will require far more memory than a tiny program running on a washing machine.

Processor

Desktop computers and smartphones contain a very fast processor known as a **CPU** (**Central Processing Unit**). By contrast, embedded systems use microcontrollers which integrate a slower processor with memory onto one chip.

More processing power will be required to run a **Graphical User Interface** (**GUI**) and software that undertakes complex processing.

---

1. State how display size is usually measured. [1]
2. Describe **one** benefit of OLED displays. [2]
3. Give **one** reason why a touch screen would be used in a tablet computer. [1]

   1. It is measured by the distance from one corner to the opposite corner of the display,[1] the diagonal size of the display.[1]
   2. They can be bent[1] which means they can be used to make folding devices (such as mobile phones).[1] They have a higher contrast[1] which makes them more appealing to the viewer.[1] They are thinner[1] which helps reduce the size of the device.[1]
   3. It allows interaction with the device/HCI without the need for a mouse or keyboard.[1] It allows for gestures to be made (such as pinch or swipe).[1]

OCR Cambridge Nationals J836 (R050) **IT – Topic 2**

2.2

SOFTWARE CONSIDERATIONS

The HCI experience of users will be affected by the **operating system** and **digital platform** they are using.

Operating systems

Most computers and digital devices need an operating system. The operating system will contain carefully thought out HCI features. Some operating systems are **proprietary** meaning that they are owned by a company and usually cannot be altered. Other operating systems may be **open source** meaning that the way they work can be altered and customised. Open source operating systems are usually free.

All the operating systems listed below are all **multi-tasking**. This means they are able to run more than one software application or app at the same time.

Windows	• Windows is a proprietary operating system commonly used in PCs, laptops and tablets. • It makes use of a Graphical User Interface.

Unix based operating systems	
Unix	• **Unix** first began in 1969 as a proprietary operating system. • Unix is rarely used today, but the way the operating system works, and many of its features are present in other Unix based operating systems. • **Unix based** systems have good multi-user support and are known for their reliability and security.
Linux	• **Linux** is a free, open source operating system. The **Linux kernel** (operating system core) is used in Ubuntu, Chrome OS and Android. • Linux users often use a **command line interface**, but a Graphical User Interface option is available and commonly used.
Ubuntu	• **Ubuntu** is an open source **Linux distribution** which comes with an easy to use GUI similar to Windows. • Commonly used on desktop and laptop computers.
Chrome OS	• **Chrome OS** is an open source **Linux distribution** developed by **Google** and used in **Chromebooks**. • A very simple to use GUI. • Designed around web-based tasks.
Android	• **Android** is an opensource Linux based operating system customised for use with a smartphone and touchscreen. • The user typically only sees an easy to use GUI.
Apple macOS	• **macOS** is based on Unix and is a proprietary operating system that features an easy to use GUI. • A **terminal window** allows text commands to be entered.
Apple iOS	• **iOS** is an intuitive proprietary operating system, based on Unix. • Features such as **gestures** and large icons help the operating system in its use on smartphones. • A similar operating system, **iPadOS** is used on Apple tablets.

HCI in operating systems

Command Line and **Graphical User Interface** (**GUI**) are two types of HCI commonly used for operating systems.

Command line

A command line interface allows the user to enter text commands using the keyboard.

- ➕ Entering commands is often faster than selecting many menu options with a GUI.
- ➕ Uses very little processing compared to GUIs.
- ➖ Commands are complicated and must be learnt by the user.
- ➖ Typically, this interface requires an advanced or technical user.
- ➖ Commands must be entered accurately for them to run.

GUI

Desktop GUIs use **windows**, **icons**, **menus** and **pointers** (**WIMP**).

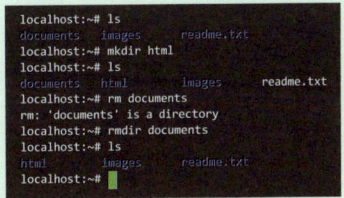

- ➕ GUI interfaces are intuitive and user friendly.
- ➕ GUIs on smartphones allow gestures such as pinch or swipe to be used.
- ➕ GUIs are usually are easy to learn and have labels and onscreen help to assist the user if they need assistance.
- ➖ These interfaces require more resources such as processing power and memory to run.
- ➖ They are often slower to respond as they require more processing to be carried out.

---

1. Explain **two** HCI features used in desktop operating systems such as Microsoft Windows and Apple macOS. [4]
2. Explain how the use of a Linux command line HCI may help an advanced user compared to a GUI. [2]
3. (a) Identify an operating system commonly used in smartphones. [1]
   (b) For the chosen operating system in part (a), explain **one** feature of an HCI that makes use of the touchscreen. [2]

---

1. Windows[1] allow each program to be resized/minimised/maximised/exited.[1] Icons[1] allow a user to easily identify software[1] and load it by double clicking.[1] Menus[1] allow users to select different options.[1] Mouse pointers[1] allow the user to see where the mouse position is currently located.[1]
2. An advanced user may be more efficient/faster[1] at entering commands compared to clicking various options in the GUI.[1]
3. (a) Android/iOS.[1]
   (b) The touchscreen allows users to: pinch the screen with two fingers[1] to zoom in/out;[1] swipe the screen[1] to alter the home screens/icon layouts[1] or to close programs;[1] touch the screen to select an icon/load an app;[1] draw on the screen[1] or stylus.[1]

2.3 HCI USE IN DIGITAL PLATFORMS

HCIs will vary depending on the digital platform they are used for. The digital platforms covered in this unit are **mobile apps**, **websites**, **spreadsheets** and **databases**.

Mobile apps

The HCI of mobile apps is usually customised for working with a touch screen. This allows for **touch gestures**. Screen designs must allow for a smaller screen size on a mobile device. **Buttons** must be larger to allow users to accurately press them.

Apps will often make use of microphones and cameras. They should also adapt to the device being held in **portrait** (vertically) or **landscape** mode (horizontally).

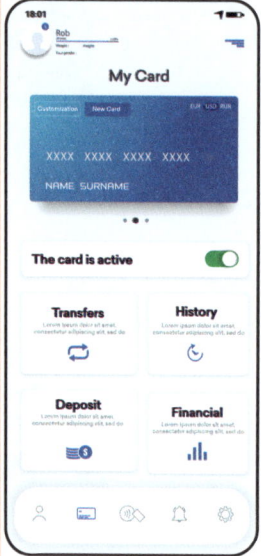

Websites

Websites are typically designed with a larger screen size than mobile apps as they may be viewed on laptop and desktop computers. Many websites will change the layout for use on mobile devices. These are known as **responsive** websites.

The use of standard elements for forms, text and video players makes the HCI easier and more consistent for the user. The **web browser** is a key part of the HCI experience.

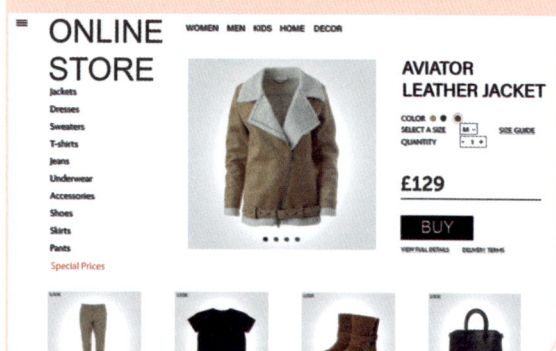

Spreadsheets

Spreadsheets use an HCI that allows the easy entry of data such as numbers and text. Cursor keys and a mouse are heavily used to navigate a spreadsheet. The HCI makes it easy for both **values** and **formulas** to be viewed.

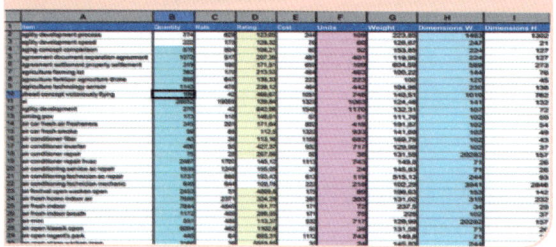

Databases

Databases may require an interface that allows programmers to enter text commands to add or search for data. Alternatively, they may use more graphical interfaces with HCI considerations such as **drop down** menus, **radio buttons** and error **pop-up** messages.

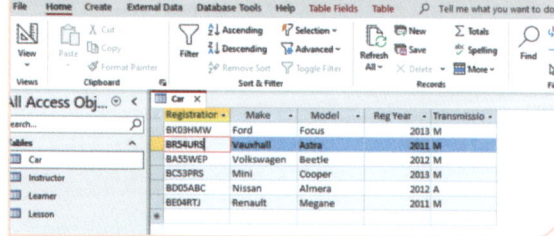

2.4 USER INTERACTION METHODS

There are many different methods available for users to interact with an HCI.

Touch

Touch is used with devices such as **buttons**, **touchscreens**, **joysticks** or **keys**.

- ⊕ A range of movements suited to games.
- ⊕ Suited to quick presses and fast reactions.
- ⊖ Some interactions aren't intuitive.
- ⊖ Accessibility issues for some disabilities.

Gestures

Gestures are used to control systems through movement. This may be on a touchscreen or with larger movements to interact with a game.

- ⊕ Intuitive – e.g. physically dragging an icon into a bin to delete it.
- ⊕ It is often more efficient. For instance, pinching to zoom, rather than clicking a +/- button.
- ⊖ Needs a touchscreen or video camera.

Mouse

Mice allow a pointer to be moved and items to be selected. **Touch pads** are often used as an alternative on laptops.

- ⊕ Accurate at moving the mouse pointer.
- ⊕ Scroll wheels and buttons are efficient and intuitive to use.
- ⊖ Users need time to get used to using them.
- ⊖ Require sufficient desk space.

Voice

Voice control is used by speaking or making noises. For instance, a **home assistant** uses voice to select music.

- ⊕ Easy to use.
- ⊕ Good for those with physical disabilities.
- ⊕ Can be used from a distance.
- ⊖ Requires a microphone.
- ⊖ Voice recognition may be inaccurate.

Keyboards

Keyboards are used for entering text. They allow for accurate and fast entry of text. **Keypads** are used on cash machines, card machines and door entry systems.

- ⊕ Intuitive for data entry.
- ⊕ Fast and accurate once a user has learnt how to type.
- ⊖ Takes a long time to learn how to use them with speed and accuracy.
- ⊖ Physically large.

> An online shop has both a mobile app and website. Explain **two** differences that are likely between the two HCIs. [4]
>
> *A menu icon will appear in the mobile app[1] as there isn't space for a full menu/navigation bar.[1] Buttons[1] will be bigger on the mobile app as they need to be selected using a finger.[1] The width of the mobile app will be less[1] as the screen is smaller, resulting in more scrolling.[1] An onscreen keyboard is required for the mobile app[1] as the mobile device doesn't have a physical keyboard.[1]*

Topic 2

EXAMINATION PRACTICE

Section A style questions

1. Two interaction methods that are used by users to interact with an HCI are touch and gestures. Identify **two** other interaction methods used with HCIs. [2]

2. Explain how voice interaction may be useful for people with physical disabilities. [2]

3. Explain why an HCI should be intuitive. [2]

4. A graphical user interface (GUI) is one type of HCI that may be used in an operating system. Name **one** other type of HCI that may be used. [1]

Section B style questions

A mobile phone manufacturer is creating a new phone for use by primary school aged children that avoids them being distracted or viewing inappropriate content. The phone will have basic features allowing parents and children to communicate with voice calls and texts. It will have no Internet connection and won't be able to install apps.

5. The phone developer is considering using a graphical user interface (GUI).
   (a) Explain why an HCI that uses a graphical user interface (GUI) may need more resources to run effectively. [4]
   (b) Explain a suitable type of display for the screen. [2]
   (c) Describe **two** advantages of using touch and gestures when children need to make voice calls or write texts. [4]
   (d) Describe **one** reason why a mouse would not be a suitable interaction method for the phone. [2]
   (e) Explain a suitable operating system that the developer could use for the phone. [2]

3.1, 3.2.1

INFORMATION AND DATA

Data is raw unprocessed text, numbers or symbols. Data on its own lacks meaning and is unorganised. For example, the number 15 is data, but what does 15 mean? 15 could be your age, the date you were born, the number of cars in a car park or the number of petals on a flower. **Information** is data with **meaning**. Information may also have a **structure** or a **context**, but it does not have to have context or structure to be useful.

$$\text{Information} = \text{Data} + \text{Meaning} + (\text{Structure}) + (\text{Context})$$

Data and information examples

Information	Data	050124
	Data + Meaning	The data is a date. (So it means 5th January 2024).
	Data + Meaning + Structure	05/01/24 (The data has been structured so that each part of the data is separated by a slash). Data is often structured into fields and rows in a database.
	Data + Meaning + Structure + Context	The United States read dates in the order month, day, year. This additional context changes the meaning of 05/01/24 to May 1st 2024.

Data types

Type	Meaning	Example
Alphanumeric	Any sequence that contains only letters and numbers.	*Username:* arichards3
Text	Any type of character including letters, numbers, spaces and symbols.	*Email address:* apatel@pgexd.com
Boolean	One of two options: TRUE/FALSE, YES/NO or 1/0.	*Hotel paid:* Yes
Date	A date in a format such as DD/MM/YYYY.	*Payment date:* 15/12/2024
Numeric	**Number** related data including: **Integer** (whole numbers) **Decimal/Real** (numbers with a decimal point) **Percentage** (numbers with a percentage symbol) **Currency** (numbers with a currency symbol and decimal point)	*Items sold:* 5 *Distance in km:* 4.31467 *Test score:* 87% *Invoice total:* £329.46

1. State what the term data means. [1]
2. Data needs meaning for it to become information. Give **two** other aspects that help give data further meaning. [2]

1. Raw unprocessed text, numbers or symbols.[1]
2. Structure,[1] context.[1]

OCR Cambridge Nationals J836 (R050) **IT – Topic 3**

3.2.2, 3.2.4

VALIDATION AND VERIFICATION

Before data is used, it often needs to be checked to make sure it is sensible or reasonable. **Validation** makes sure data meets certain requirements or rules. **Verification** makes sure data is accurate or consistent.

Verification

There are two types of verification you need to be aware of.

Manual Checking

1. The data is entered and stored.
2. The data is then **manually checked** by proofreading it against the original data.
3. If the data doesn't match then it is either rejected or corrected.

Double Entry verification

1. The same information is typed in twice and compared.
2. If the data does not match exactly, it is rejected.
3. The user then needs to enter the data again.

Double entry verification is commonly used when choosing passwords. This helps to make sure that the user hasn't made an error when they first enter their password.

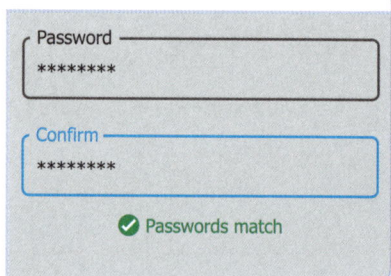

When a customer opens a bank account they give the bank their driving licence. They then hand in a form which contains their name and date of birth. The bank employee then verifies their identification (ID).

(a) Describe how manual checking has been used as a verification method. [2]

(b) Explain how double entry verification could be used as a verification method when the customer opens their bank account. [2]

(a) The bank employee compares the name and date of birth on the form with the same details on the driving licence.[1] If the details match then the ID is verified.[1] Otherwise the application is rejected (due to failing the ID manual check).

(b) They could be asked to enter their password/email address/PIN twice[1] with both passwords/email addresses/PINs needing to match to be verified.[1]

3.2.3

DATA VALIDATION TOOLS

Validation tool	Details about the validation tool	Example
Data type check	• Checks data match a particular data type	Enter length: 3 metres Error: Length must be an integer
Format check	• Checks data matches a given format once it has been entered.	Postcode: SW1A 1DEF Error: Incorrect postcode format
Input mask	• An input mask allows for certain characters to be entered in spaces provided.	Phone: (...) _____ (Area code followed by number)
Length check	• Ensures the length of text fields are a certain number of characters long or between a maximum or minimum length.	Password: 4vQE76&p Error: Password must be at least 10 characters long.
Limited choice / Lookup	• Limit data entry to pre-defined options. • **Drop down lists**, **checkboxes**, **toggle buttons**, **radio buttons**. • Lookups choose options from a list of valid items. • Reduces the chance of spelling mistakes.	**Drop Down list** — Country: France, Spain, UK, USA **Checkboxes** — ☑ Bus, ☐ Taxi, ☑ Car **Toggle button** — On **Radio buttons** — ○ Small, ● Medium, ○ Large
Presence check	• Data must be entered. • Leaving the item blank creates an error.	City*: Error: You must enter a city
Range check	• Checks that a number entered is between preset minimum and maximum values.	Age: 12 Error: You need to be 13+.

1. Describe the difference between validation and verification. [2]
2. A website requires people to enter their date of birth.
   Explain **one** validation tool that could be used by the website on this data. [2]

> 1. Verification checks that data is accurate/consistent/checked against an alternative,[1] validation ensures data meets certain requirements/rules.[1]
> 2. An input mask[1] could provide spaces for the user to enter the numbers[1] and add slashes/hyphens to separate the parts of the date.[1]
> A range check[1] could make sure the date allows the user to be an appropriate age / such as 13–120 for a social media network.
> A presence check[1] would make sure the user entered the date.[1]

Be careful. Validation ensures that data meets sensible criteria. For example, an incorrect date of birth may still meet the criteria even though it is wrong.

3.3

DATA COLLECTION METHODS

When data is collected there are two broad types of method used. **Primary data collection** involves a person or organisation collecting the data themselves. **Secondary data collection** involves using data that another person or organisation has collected.

Primary data collection methods

Email

An **email** is sent which contains a link to an online form. The users complete the form online and data is collected into a database or spreadsheet.

- ➕ Quick to send.
- ➕ May be tracked to see if the recipient opens the email.
- ➕ May be sent all over the world in seconds.
- ➕ Electronic responses are automatically collated.
- ➖ Needs digital skills to create.
- ➖ The email may end up in junk or spam folders.
- ➖ Recipient needs an email address and basic digital skills to respond.
- ➖ Easy to ignore or forget about.

Interview

In **interviews**, an interviewer asks questions and an interviewee answers them. They take place face to face, on the phone or by video conference.

- ➕ The interviewer is able to see people's reactions.
- ➕ Follow up questions may be asked.
- ➕ More personal.
- ➕ More considered responses.
- ➖ Hard to arrange meetings and suitable times.
- ➖ Takes longer to do.
- ➖ May need to hand write the responses.
- ➖ The data will take time to collate and analyse.

Online questionnaire and survey

Questionnaires contain questions that may be delivered online or on paper.

- ➕ Paper questionnaires don't need computers to complete.
- ➕ Paper questionnaires may have a higher response rate if someone is waiting to collect them in.
- ➖ Paper surveys are more expensive to produce.
- ➖ Digital surveys require digital devices to complete.
- ➖ Collecting paper based surveys later is awkward.
- ➖ Handwriting may be difficult to read with paper questionnaires.
- ➖ Effective digital surveys take a while to design.

1. A local dentist wants to find out how happy their patients are with their service. They decide to use an online questionnaire.

   Explain **two** advantages of using an online questionnaire rather than paper questionnaires left at reception. [4]

   1. Data will be automatically collected/collated[1] saving time.[1] Patients won't forget to hand them in/lose them[1] helping to increase response rates.[1] Patients will be able to fill them in privately[1] so may be more honest.[1]

ClearRevise

Secondary data collection methods

Books / Magazines

Books may compile statistics, data or other information.

➕ They tend to be reliable as they are checked by editors and proof-readers.

➕ They are usually available both online and in print.

➖ They may be out of date as physical books cannot be updated once printed.

➖ It takes time to publish a book or magazine so they cannot be used as data collection methods of very recent events.

➖ The author or publisher could be biased.

Websites

Websites contain huge amounts of information. It is easy and cheap to publish on the web.

➕ There are a huge number of websites available.

➕ Quick to search.

➕ Access to data sources across the world.

➖ Not always reliable.

➖ Need to know how to search the Internet effectively.

➖ There are lots of unreliable and biased websites and identifying the owner/author is often difficult.

Government Statistics

Government statistics are produced by many public organisations to report information such as hospital or school performance.

➕ Usually viewed as a reliable source.

➕ Has a wide range of data on it, from health, population data and local information.

➕ The statistics are often free to download.

➖ Datasets may be large and contain lots of unnecessary information.

➖ The government could be biased in the data it releases or how they process it.

---

2. Give **two** drawbacks of using interviews for data collection. [2]
3. State **one** benefit of using government statistics. [1]
4. A travel agent is researching interesting day trips and information to include in a travel brochure.
   Give **two** drawbacks of using websites to find the information they require. [2]

---

2. Answers must be written down.[1] They are hard to arrange.[1] They take longer to carry out.[1] Both people must be available.[1] They need to be collated/take more time to analyse.[1]

3. They are usually a reliable source,[1] provide a wide range of data,[1] often updated regularly,[1] often free to use.[1]

4. The information such as times/costs may not always be reliable.[1] They will need good searching skills to find the information.[1] The information may be unreliable/biased due to the experience of a particular person writing it.[1]

3.4

STORAGE OF COLLECTED DATA

Data needs to be **stored** somewhere in order for it to be processed or viewed later. It may be stored digitally in a **physical location** either using an **internal storage device** or an **external storage device**. Alternatively, data may be stored in the **cloud**.

Network drives

Network drives are located on **file servers**. These allow files to be shared across a **network**. They allow access to a user's files from any computer on the network. It is vital that these are backed up to prevent loss of many user's data should they fail.

Network drives use hard disk drives or solid state drives to store their data.

Internal storage devices

Internal storage devices are used within a computer or device. A primary hard drive may be either a **magnetic hard disk drive** (**HDD**) or a **solid state drive** (**SSD**). Solid state drives are faster to access data from, are smaller, making them more portable and they require less power. However, magnetic hard disk drives are cheaper and therefore used for larger storage volumes.

Internal hard disk drive (HDD)	Internal solid state drive (SSD)

Cloud storage

Cloud storage is a term used for remote storage that is accessed via the **Internet**. This type of storage is a **logical location** rather than a physical one. The cloud storage provider could physically store the data anywhere in the world. The user just knows it is somewhere in the **cloud** on the Internet. The user is able to access the data from anywhere they have access to the Internet.

Cloud storage examples include:
- Microsoft OneDrive
- Google Drive
- DropBox

Cloud storage also allows users to share their files with others by using **hyperlinks**. This means they are able to collaborate with friends or colleagues and access the same file without needing to email it. The shared files may require a username and password to access them.

Data stored on the cloud will usually be **backed up** and stored very **securely** on behalf of the customer. This makes cloud storage more convenient. However, the cloud storage company is able to access any data stored with it and the customer is normally unable to verify that backup procedures and security are sufficient. As such, they must trust the cloud service provider that they use.

1. Describe **one** advantage of an employee on a business trip using the cloud for file storage. [2]

   1. They can access their data from anywhere[1] by using any available Internet connection.[1] They can share data easily with other colleagues[1] by giving them links to their files.[1]

External storage devices

External storage devices are located outside the computer or device.

Portable external hard disk drives (HDD) and portable solid state drives (SSD)

These drives are similar to the internal drives used in computers, however, the drives are placed inside a protective case and are connected using a cable.

A **portable wireless drive** works in the same way, but is powered by a battery and is able to share data using a **Bluetooth** or **Wi-Fi** connection.

- ➕ Easy to install.
- ➕ Larger capacity than flash drives.
- ➕ May be used to back up an internal drive.
- ➕ Wireless version is convenient to use and may allow multiple connections at once.
- ➖ Easy to lose.
- ➖ Need a cable to connect to a computer or device.
- ➖ Still needs to be backed up if this is the only device used to store data.
- ➖ May be stolen.
- ➖ May break if dropped.

Network-attached storage (NAS) device

These devices contain a storage drive, processor and RAM. They store files and may allow file sharing. They are similar to file servers, but more suitable for home users and smaller organisations. They must be connected to a router.

- ➕ Allows the sharing of files in a central location.
- ➕ The owner remains in control of the data.
- ➖ If the drive fails, more users may be affected.
- ➖ Vital that the device is backed up.
- ➖ Harder to setup and configure.
- ➖ Needs network access.

Portable USB flash drives

These are small solid state drives that connect to the USB port of a computer.

- ➕ Small size makes them easily portable.
- ➕ Compatible with any computer with a USB port.
- ➕ Cheap.
- ➖ Easy to lose or steal.
- ➖ May get easily broken.
- ➖ Very important .to backup given the risk of being lost, stolen or broken.

---

2. Describe why a network-attached storage (NAS) device would benefit a small company with an internal network. [3]
3. Give **two** issues a company may have when using a portable external hard disk drive. [2]

2. All the company files could be stored in one location[1] which would allow them to be accessed from any of the computers on the network.[1] The use of a central location makes it easier to back up.[1] Files may be set for sharing[1] which would allow others outside the organisation to access them.[1] The NAS is owned and controlled by the company[1] which means they are in total control of the data.[1]

3. The drive could easily be lost / stolen.[1] It would be easy to break by dropping it.[1] It needs a cable to connect it to a computer.[1] Someone will need to remember to carry out a backup.[1] A presence check[1] would make sure the user entered the date.[1]

3.5

TESTING

Testing is important to ensure that a solution to a problem is error free, meets the needs of the users and meets the requirements of the project.

Advantages and disadvantages of testing

➕ Advantages	➖ Disadvantages
• Demonstrates that the software works. • Ensures that requirements are met. • May find unexpected bugs or issues. • Helps to make software more reliable. • Helps to improve the user experience.	• Building thorough test plans takes a long time. • The testing itself takes a lot of time. • It is not usually possible to test all possibilities. • Testing and fixing errors takes a long time and may delay a project. • Correcting errors may cause new errors to occur.

Test data

There are three types of data which are used when testing programs and software:
- **Extreme**: Data which is right at the boundary of being valid. Also called boundary test data.
- **Invalid** (Erroneous): Data that causes an error.
- **Valid**: Data which should be accepted.

Examples of test data

Imagine a website which requires a user to enter their age as a whole number. The website should allow the age to be between 13 and 120 inclusive. The following would be examples of tests that could be performed.

Type of test data	Test data examples	Reason
Extreme	13, 120	These numbers are on the boundary of valid data.
Invalid	−1, 12, 121, "fifty"	Numbers outside of the range would be invalid data. Text is also invalid.
Valid	13 to 120	Any numbers from 13 to 120 would be valid data. Typically, a programmer would test a few numbers to see that they work correctly. Remember that 13 and 120 are both valid and extreme data as they are at the boundary of valid data.

Types of testing

Technical testing

Technical testing checks that the software or system works as intended. It checks the functionality of a system and may also check that software works on different devices.

Technical tests include:
- Specific data entry testing (extreme/valid/erroneous)
- Menu functionality
- Navigation including hyperlinks
- Forms and buttons
- Speed or response times of the system
- Reliability.

⊕ Ensures errors and bugs are fixed and the software works as intended.

⊕ Gives confidence that the system meets the needs of the users.

⊕ Demonstrates that it works on different devices if appropriate.

⊖ Time consuming to test everything.

⊖ Real users may use the system in ways that haven't been thought of.

⊖ May need a lot of testers which will increase costs.

User testing

User testing gets users to test the system. Testers will give users problems to solve with the software. This may be given on a test plan or alternatively the developer may watch how .they use the system. This type of testing may be carried out during or at the end of development.

User tests include:
- Entering data and checking that validation works correctly for real user inputs
- Interactions with the system such as using menus, forms and buttons are intuitive
- Usability testing to see how intuitive the system is to use.

⊕ Gives 'real life' feedback.

⊕ Users may use the system in different ways to a technical tester.

⊕ Allows developers to watch and follow users using the system.

⊕ Builds confidence with users.

⊕ Gets a broad range of feedback.

⊖ Need a range of users to test it.

⊖ Very time consuming to carry out.

⊖ Poor experiences may put people off using the system in the future.

---

1. Give **two** advantages of testing a solution. [2]
2. One type of test data is valid data. State **two** other types of test data. [2]
3. Describe how user testing would be carried out for new software. [3]

   1. Demonstrates the software works,[1] checks it meets requirements,[1] finds bugs/errors/issues,[1] increases reliability,[1] helps improve the user experience.[1]
   2. Extreme,[1] Invalid.[1]
   3. A test user uses the system[1] to complete tasks given on a test plan[1] whilst they are recorded/watched by the developer.[1] This occurs during or at the end of the test process[1] and helps to inform further development.[1]

EXAMINATION PRACTICE

Section A style questions

1. Explain the difference between data and information. [2]

2. A software program makes use of validation and verification tools with data entered in a form.
   - (a) Give **one** reason why validation tools should be used when capturing data. [1]
   - (b) Describe how double entry verification works. [3]
   - (c) State **three** data validation tools that the software program could use. [3]

Section B style questions

A local council wants to have exceptional beaches ready for next summer's tourist season. They have had reports from the public that litter is currently an issue.

3. The local council wants to research the public's views of the local beaches.
   - (a) Describe how **one** type of primary research could be used by the council. [3]
   - (b) Describe how **one** type of secondary research could be used by the council. [3]

4. The local council wants to electronically store the data collected and share it with other organisations in the country.
   - (a) Give **two** advantages of using the Cloud to store and share the data. [2]
   - (b) Give **two** disadvantages of using a portable hard disk drive (HDD) to share the data. [2]
   - (c) Name an alternative type of external storage that the council could use to share the information. [1]

5. The local council will be updating their website. They will test the website before they make it live for the public to visit.
   - (a) Explain how user testing could help when testing the website. [2]

   One webpage on the website asks the user to enter the age of a child to suggest suitable attractions near beaches. The age will be entered in years and valid ages will be between 1 and 15.
   - (b) Give **one** data type that could be used for the age. [1]
   - (c) Describe how extreme test data could be used to test this part of the website. [2]

4
CYBER-SECURITY AND LEGISLATION

Cybersecurity covers how electronic data, devices and networks are protected from the many **threats** that exist. Threats may be attempts to gain unauthorised access to computer systems and data. Other threats include damaging or disrupting computer systems or networks.

Threats can be reduced by effective **prevention measures**. Prevention measures aim to reduce the risk of threats. No prevention measure is 100% effective against all threats.

Legislation adds legal means to deter people from committing cybercrime.

Types of hacker

Hackers are people who try to gain **unauthorised access** to data, computer systems or networks. Some hackers may be given permission to attempt to hack a computer system or network.

Black hat

Black hat hackers try to gain access to computer systems or networks with **malicious intent**. This means that they aim to cause damage or steal information. They will often work for organised crime with money often being a key motivation.

This type of hacking is **illegal**.

Grey hat

Grey hat hackers try to gain access to computer systems or networks just as black hat hackers do. However, they do not have malicious intent. Their motivation is usually for a challenge or fun. They may also enjoy discovering **vulnerabilities**, however, they don't **exploit** these when found.

Grey hat hackers do not have permission to carry out their hacking attempts. This makes this type of hacking **illegal**.

White hat

White hat hackers try to find vulnerabilities in computer systems but they have the owner's agreement to do this and are often paid. This type of hacker is motivated by either challenge or payment from the company.

Companies employ white hat hackers to try and break into their systems. This helps them to identify weaknesses and make the system more secure.

As white hat hackers have the owner's permission, this type of hacking is **legal**.

Explain an appropriate type of hacker for a company to use to check the security of their computer systems. [2]

A white hat hacker[1] is appropriate as they only take actions they have been permitted to do[1] and this type of hacking is legal.[1]

OCR Cambridge Nationals J836 (R050) IT – Topic 4

4.1

THREATS

Threats to computer systems include **hacking**, **denial of service**, **malware** and **social engineering**.

Denial of Service (DoS)

In a **denial of service** attack, one computer that is owned by, or under control of a hacker will **flood** a server with requests. The server spends so much time dealing with these that other users are unable to use the service. This type of attack is often used to prevent access to a website.

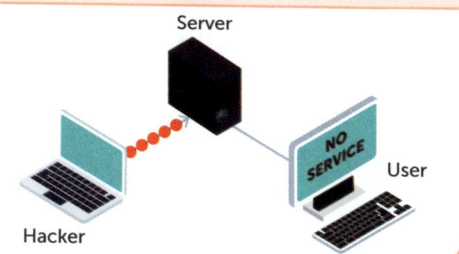

Malware

Malware is a term used to describe malicious software which, when run, will harm a computer system or data.

Adware

Adware shows adverts, such as pop-ups which are often intrusive or irritating. It may be hard to turn these off.

Ransomware

Ransomware encrypts data on a device to prevent access. A fee is usually demanded to decrypt the device.

Spyware

Spyware watches and records activity on a device such as websites visited or keystrokes pressed.

Worm

Worms are similar to viruses, however, they replicate on **their own** without the user needing to run them.

Botnet

In a **botnet**, a hacker will control many **zombie** devices across the Internet without the users of the devices knowing. Botnets are often used in distributed denial of service attacks.

Trojan horse

Trojan horses are malware that hides inside standard software. When the software is run, the malicious activity also runs. A backdoor may be created through a Trojan horse. This allows the hacker access to the computer.

Virus

Viruses are malicious programs that damage files or corrupt systems. They replicate and spread to other devices by tricking the end user into running them.

Social engineering

Social engineering uses manipulation of people to get around security in computer systems. For instance, social engineering could be used to trick someone into giving away their password.

Type	What is it?
Baiting	**Baiting** is where someone tries to get users to give confidential information by offering a reward such as bitcoin or free goods. Once the information is given, the reward will not be given.
Phishing	**Phishing** makes use of fake emails or websites to trick a user into entering sensitive data/information. For example: replicating a bank website to look like it's official. A user then enters their banking details into the fake website.
Pretexting	**Pretexting** gives the user a pretend scenario which helps persuade them to give confidential information or take an action such as giving access to a computer system. Pretexting could involve a phone call pretending to be from a software supplier and asking the user to install software that contains a backdoor, allowing unauthorised access.
Quid pro quo	**Quid pro quo** is a Latin phrase that means 'something for something'. In a quid pro quo attack an attacker will offer a benefit or service. For instance, they could offer to remove a virus from the user's computer. To receive the benefit, the victim must take an action such as installing malware or giving away confidential information.
Scareware	**Scareware** scares the user into buying or downloading something unnecessary. For instance, a pop-up on a website could claim the computer has a virus and needs to install a virus checker. This virus checker would then contain malware such as ransomware.
Shoulder surfing	An attacker will look over a victim's shoulder to gain confidential information such as a computer password or PIN (Personal Identification Number) entered into a card machine.

1. State the meaning of the term malware. [1]
2. Give **one** difference between a virus and a worm. [1]
3. Describe how a denial of service attack prevents a website from working. [3]
4. (a) A victim receives a phone call from someone pretending to be from their bank. They say there has been a problem and they just need to verify their bank account number to authorise them to deal with it.
   Name **two** types of social engineering that have been used in this scenario. [2]
   (b) A café allows people to use laptops to do their work during the day.
   Describe how shoulder surfing could be a threat for their customers. [3]

1. Malicious software[1] – software which aims to cause harm to a computer system or data.[1]
2. A worm can self-replicate / spread by itself,[1] a virus needs a person to copy/open/run the program for it to be able to replicate.[1]
3. A hacker will cause their computer / a computer they control (such as a zombie computer)[1] to flood the webserver[1] with many (pointless) requests for the website.[1] The web server won't be able to process these fast enough[1] resulting in the website not being available for other users.[1]
4. (a) Pretexting[1] and Quid Pro Quo.[1] (The pretext part is that they work for the bank. The quid pro quo part is that they will fix a problem if the victim gives their bank account number.)
   (b) Shoulder surfing is looking over someone's shoulder at what they type[1] in order to gain personal information.[1] This would be particularly easy to do in a café as the attacker would have a good reason to walk past the customer[1] or sit at a nearby table where they can see the keyboard.[1]

4.2

IMPACTS OF CYBER-SECURITY ATTACKS

When a cyber-security attack occurs, there are many impacts that this may have on individuals or organisations. Key impacts are given below which may lead to loss of money and reputation. Safety systems may also be compromised.

Impacts on individuals and organisations

Data destruction

Viruses, worms and Trojan horses may result in **data destruction** where any data stored on internal or external drives is **deleted**. A hacker is also able to destroy data if they gain access to a computer system.

Data manipulation

Data manipulation is any alternative processing of data to meet the needs of an attacker. For instance, the contents of a company report could be altered to give alternative conclusions.

Denial of Service (DoS)

Denial of Service attacks prevent **authorised users** from using a service such as accessing a website.

Data modification

Data modification changes the original data. For instance, the amount of money in a hacker's bank account could be increased. A hacker may make other modifications to system files or passwords allowing them further access to the computer system.

Data theft

Data theft could occur if an external drive or storage media is in transit, such as being left in a car, bus or train. Alternatively, the data may be stolen from computer systems at rest. A drive may be physically removed from a laptop, computer or server. Data may also be **copied remotely** through a network. Use of **encryption** decreases the risk of the data being understood if it is stolen, however, **locks** and other security measures are equally important.

Identity theft

Personal information may be stolen through unauthorised access to a system. It could also be found in public sources such as social media accounts.

Identity theft is where a hacker or criminal uses personal information to pretend to be someone else. They will typically use their new identity to make online orders, steal money or take out loans.

A hacker has gained access to an online shopping website resulting in data on the website being manipulated.

Describe **two** more impacts the company may experience as a result of the hacker's actions. [4]

The hacker may have destroyed/deleted data[1] such as database entries/web pages/graphics.[1] They may have modified permissions/rights[1] to give themselves further access[1] or to alter how the website processes data.[1] The hacker may have stolen confidential information[1] such as company sales data.[1] They may have stolen personal data[1] which could lead to identity theft of customers.[1] The company may have suffered reputational damage[1] or have other financial impacts.[1]

PREVENTION MEASURES

There are many ways that help to prevent cybersecurity threats from being successful. These fall into three broad categories, **physical prevention measures**, **logical prevention measures** and **secure destruction of data**.

Physical prevention measures

Physical prevention measures are physical hardware or devices which prevent access to computer systems or data. The following are the devices you need to be aware of.

Biometric devices

Biometric devices work on a person's physical characteristics. **Fingerprints**, **facial recognition** and **eye scans** (of the iris or retina) are commonly used. Biometrics are also often used to gain access to portable devices such as smartphones.

Keypads

Keypads are often used to secure doors or safes. They are also used on smartphones and cash registers. A code is entered to unlock the room or device.

Radio-frequency identification (RFID)

RFID is used in door entry cards and fobs. The card is held close to the device causing a door to be unlocked.

Firewalls

Firewalls sit between an external and internal network. A physical firewall is a hardware device that sits between the two networks. Data is sent between networks in **packets**. The firewall **blocks** harmful packets coming from the Internet so they don't enter the internal network.

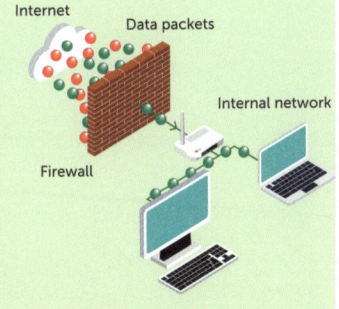

Secure backups

Backups of data are important as they allow data to be **restored** if it is damaged or deleted.

It is important that backups are stored securely, such as in a locked room with other security features in a remote location.

Logical prevention measures

Logical prevention measures use software to prevent threats.

Access rights and permissions

Access to files, apps, software and settings is restricted to different users.

Two-factor authentication (2FA)

Factors are either something a user is (such as biometric information), something they have (such as a key fob) or something they know (such as a password). **Two factor authentication** makes use of two different security methods, each using a different factor.

Anti-virus / anti-malware software

Anti-virus software scans programs or files to check that they don't contain **signatures** that indicate they are **infected**. If they do, they will be **quarantined** and prevented from running.

Encryption

Encryption involves **encoding information** so that it cannot be understood without a **security key** or **password**. Information stored on hard drives or sent over a network should always be **encrypted**.

Software firewalls

Software firewalls work in the same was as hardware firewalls except they are implemented using software running on a computer. They typically only protect one computer or device.

Secure backups

Data may be **backed up** on the same device. This allows files to quickly be recovered if a user accidentally deletes them. It is important that files are also backed up physically and stored in a separate, **remote location**.

Usernames and passwords

Usernames and passwords are commonly used when logging into computer systems or websites. The username identifies the user whilst the password is memorised so that access is only given if the user knows it .

---

1. Explain how a banking app could make use of two-factor authentication before money is transferred to another bank account. [2]
2. Describe how anti-virus software helps to prevent computer viruses. [3]

   1. *The user could log in with their password (something they know).[1] They could receive a text message with a code (the phone is something they have).[1] They could use a fingerprint scanner on their device (something they are).[1]*
   2. *The software scans the programs/files[1] when downloaded/when opened/as a full drive scan.[1] If a signature of a virus is found[1] the file/program is quarantined[1] preventing it from being opened/run.[1]*

Secure destruction of data

When you **delete** files or **format** a hard drive, the information often remains stored on the disk. The operating system simply allocates the area as free space. Technical users and hackers are often able to **recover** the information. It is therefore crucial that data erasure occurs before any storage media or drives are **disposed** of or re-used.

Data sanitisation

Data sanitisation involves writing over the area on the disk where the data was stored with random data. This **erases** the original data preventing it from being recovered. The data sanitisation procedure could be applied to the entire drive to wipe everything from it.

```
Confidential document.
Make sure to use data
sanitisation to erase.
```
↓
```
iJfPQK9oxzsRwoO2ezWx97fd
78P5GaXWYBNj3LYWobc13IGX
LkyLfxlyMD0szSq8eXfIn86L
```

Physical destruction

Drives may be physically destroyed so that data on them cannot be read. In the case of solid state devices or removable media such as CDs or DVDs, they may be **shredded**.

Magnetic wipe

A powerful magnet is able to perform a **magnetic wipe** on hard disk drives (HDDs). This is also known as **degaussing** a hard drive. This technique won't work with solid state drives, USB flash drives or media such as CDs or DVDs as they do not rely on magnetism to store their data.

3. A company has a hard drive in an old computer. They want to reuse this hard drive in a new computer.
   (a) The company will use data sanitisation on the drive before moving it to the new computer. Describe how data sanitisation works. [2]
   (b) Explain why physical destruction would not be a suitable method of destroying the data on the drive in this situation. [1]

   3. (a) The data on the drive is erased[1] by writing random data/binary across the entire drive.[1]
      (b) Physical destruction of the drive makes it completely unusable.[1] In this scenario, the drive is needed for a new computer, which wouldn't be possible if it had been physically destroyed.[1]

4.4 LEGISLATION RELATED TO IT SYSTEMS

Legislation is the laws that are made by the government. Another name for these laws is Acts. There are many Acts that apply to the use of IT systems and help to reduce the risk of threats.

Computer Misuse Act

The aim of the **Computer Misuse Act** is to protect computer systems and the data stored on them. It focuses on hacking, data theft and misuse of digital devices.

The following is made illegal	Example of an illegal action
Unauthorised access to **computer material** (systems and data).	Logging in to someone else's account on a computer system without their permission.
Unauthorised access in order to commit further crimes.	Hacking into a computer to find out personal data, which is then used for fraud.
Unauthorised actions which damage or weaken a computer system.	Deliberately sending a virus or Trojan horse to a computer.
Unauthorised actions that cause or create the risk of serious damage.	Hacking a railway crossing so that it doesn't function correctly.
Making, supplying or obtaining anything for use in computer misuse.	Downloading a program which attempts to find a user's password to a computer system.

Copyright, Designs and Patents Act

The **Copyright, Designs and Patents act** protects the **Intellectual Property** (**IP**) of creators of works such as books, software, music, films, photographs and illustrations.

The © copyright symbol is used along with the year the **work** was created.

How long UK copyright lasts:	Examples of illegal actions:
<ul><li>Books and scripts: 70 years after the death of the author.</li><li>Sound/music recordings: 70 years after publication.</li><li>Films: 70 years after the last director, scriptwriter or composer dies.</li></ul>	<ul><li>Installing gaming software without permission/payment.</li><li>Sharing a music track with a friend without permission/payment.</li><li>Recording a film on a phone then sharing with friends.</li></ul>

Data Protection Act

The **Data Protection Act** (**DPA**) aims to make dealing with personal data safe, secure and fair. Any organisation that stores or processes personal data will need to make sure their storage and **processing** of the data is compliant with the DPA. **Personal data** is that which identifies and describes an individual.

The **data controller** is the person or company who will collect and store the personal data. The people whose data is used are known as **data subjects**.

Data subjects are able to request to see what data is stored about themselves by making a **subject access request** (**SAR**).

The key principles of the DPA are that information must be:
- used fairly, lawfully and transparently
- used for explicitly specified purposes only
- used in a way that is adequate, relevant and limited to only what is necessary
- kept accurate and up to date
- not kept any longer than necessary
- is held with appropriate security and protections against unauthorised processing, access, loss, destruction or damage.

Freedom of Information Act

The **Freedom of Information Act** (**FOIA**) allows people to find out any official information held by a public organisation such as the government, councils or hospitals. The aim is to make the actions of public organisations more transparent.

People have a right to access the information unless there is a good reason for it to be withheld.

A **freedom of information request** must be made in writing. The requests are usually free to make, although small expenses such as postage may apply. A reply, with the information, should be made within 20 working days of the request.

---

1. A student oversees a friend's password as they enter it on a school computer. Later they log in as if they were their friend. Whilst logged in they find a music track from 1995 which they make a copy of. [2]
   Identify **two** ways that the student has broken the law.
2. A cinema asks the public to enter their card details when buying tickets online. A year later, the company has these card details stolen in a cyber-security attack.
   Give **one** way the cinema may have breached the data protection act. [1]

> 1. Computer Misuse Act.[1] (By logging in, they gain unauthorised access to computer material which is prohibited by the act).
>   Copyright, Designs and Patents Act.[1] (Copying works (music in this case) that are currently in copyright is not allowed until 70 years after it has been published.)
> 2. The card details were probably kept longer than necessary.[1] The card details did not have appropriate security/protections against unauthorised access.[1]

4.4

HEALTH AND SAFETY AT WORK ACT

Organisations and companies are responsible for making sure that their workplaces and employees are safe when working, and that risks to health are kept to a minimum. The **Health and Safety at Work Act** covers all types of workplaces.

Health and Safety (Display Screen Equipment) Regulations

One specific area that applies to computer use is **The Health and Safety (Display Screen Equipment) Regulations**. The term DSE is often used to mean Display Screen Equipment.

If an organisation expects staff to work with computers they must:
- carry out a DSE workstation assessment.
- reduce risks, such as by allowing workers to take breaks from looking at screens.
- provide an eye test if a worker asks for one.
- provide training and information on how to safely use equipment.

Display: This should be at eye-level.

Arms: Relax shoulders. Forearms should be parallel to the floor. There should be minimal bend at the wrist.

Chair: This should be height adjustable with a backrest and arm rests.

Thighs: These should be parallel to the floor.

Feet: Parallel to the floor or on a footrest.

A small company designs and installs swimming pools. They have considered the health and safety required for any building works, and are about to consider the Health and Safety requirements for using a laptop computer in the office.

Give **three** ways the company could reduce the risks of using the laptop computer. [3]

Allow/train user to take regular breaks when looking at the screen.[1]
Give training/information on how to safely use the laptop.[1]
Provide an eye test if a worker asks for one.[1]
Make sure the laptop is positioned at eye level/slightly below eye level/use a laptop stand to boost the height of the laptop.[1]
Use a height adjustable seat[1] *with five wheels/feet for stability.*[1]
Provide a foot rest if needed.[1]
Allow any other advice about the posture of the user.

Topic 4

EXAMINATION PRACTICE

Section A style questions

1. One type of malware is known as ransomware.
   (a) Describe how ransomware works. [3]
   (b) Identify the type of malware that is able to self-replicate. [1]
   (c) Identify **two** other types of malware which are a threat to computer systems. [2]

2. Explain the term white hat hacking. [2]

3. Identify the threat which attempts to stop users from accessing services by overloading a network system with requests [1]

Section B style questions

Osprey is a new bank which only operates online via an app. The app allows users to save or transfer money and also pay bills or make cardless payments. The bank is currently looking at the security of their new app.

4. The app currently asks users to enter their username and password to log in. A security review has revealed that this method alone will not be secure enough for the bank. In addition, they propose to use two-factor authentication (2FA).
   (a) Explain how two factor authentication will make it more secure. [2]

   At the bank's head office, employees must swipe an RFID card to open a barrier at reception. This reduces the chance of an unauthorised person from entering the building.

   (b) Name **two** other physical prevention devices that could be used to prevent someone from physically gaining access to the building. [2]

   Bank staff work on desktop computers inside head office. Junior staff will answer customer queries and be able to review information in their accounts. More senior staff will be able to transfer money between accounts.

   (c) Describe how the use of access rights and permissions will help to improve security. [2]
   (d) Bank staff need to be able to send confidential information to different offices via the Internet. Explain how encryption makes this more secure. [2]
   (e) The bank makes use of a physical firewall to protect their internal network. Describe how the firewall helps to protect the network. [2]
   (f) A bank employee notices a famous customer's name. They tell a friend how much money the customer has in their bank account.
   Identify the legislation that the employee has breached. [1]

5.1

TYPES OF DIGITAL COMMUNICATIONS

Digital communications are any methods of communication between people that take place electronically. The key types of communications to be aware of are given below.

Leaflets

Leaflets are small, flat or folded sheets to provide information. Designing leaflets is harder than basic word-processed documents such as reports, however, the use of **templates** makes it easier. They are typically printed and can be posted or emailed.

Newsletters

Newsletters may be one or more pages. They are often printed. Alternatively, electronic versions may be downloaded or emailed. Newsletters are used for communicating with customers and also employees as they look attractive. As the design is time-consuming to create, templates are often used.

Presentations

Presentations are used to show a series of **slides** on a screen. They are used to communicate ideas in lessons, lectures and training. They are typically quick to create and allow text, images and video to be shown on slides. Whilst slides can be printed, this is not as effective as the digital presentation and cannot contain media such as video.

Infographics

Infographics make **statistics** and **facts** appealing and attractive to read. However, they take time to create and need good design skills to be useful and attractive.

Reports

Reports are often used by organisations to communicate text and graphics. They may be well presented, but typically they are quickly made on a word-processor. **Charts** are often added to make statistics easier to understand.

Collaboration tools

Collaboration tools allow people to work on the same **documents** at the same time.

Meeting software allows users to have discussions using **video conferencing**. Presentations may be given and **screen sharing** allows users demonstrate or take control of another computer. This is useful for meetings, training and product demonstrations. Collaborative tools make people more efficient, however there is usually an additional cost to the software.

Audio

Audio includes recordings of phone conversations, meetings, podcasts and radio broadcasts and adverts. These all allow original speech to be shared and replayed. It usually takes longer to find information when in audio form than with written text.

Voice over Internet Protocol (VoIP)

Voice over Internet Protocol is a way that telephone calls can be delivered via the Internet. This method of making calls is often far cheaper than standard phone calls, however, it relies on a high quality Internet connection.

Websites

Websites allow organisations to share information, buy products or services and book tickets or holidays. An organisation will have full control over their website, but if there are lots of features, they can be costly to develop and maintain.

Video

Video is a great way for companies and influencers to communicate their products and services with customers. It is also commonly used for training and learning.

Videos can be paused, slowed or replayed which helps customers or learners to see or learn more. It requires a relatively fast connection.

Social media

Social media allows users to post messages to friends, groups or privately. Companies often use it to get feedback from customers or to help users. Social media has a very large user base, however the company usually has to use a third party system which they are not in control of.

1. Identify **two** methods of communication that a company could use to report sales figures to investors. The methods need to combine infographics, charts and text. [2]
2. Give **one** way a company could use audio to communicate with customers. [1]
3. Give **one** drawback of using a paper newsletter to communicate with customers. [1]

---

1. Report,[1] presentation,[1] website,[1] newsletter.[1]
2. A radio advert.[1] A podcast around products/services they sell.[1]
3. It will be expensive to print/post.[1] It will take time/cost to create a good design.[1]

5.2 SOFTWARE

Organisations typically use **standard office applications** including **word processors**, **spreadsheets**, **databases** and **presentation software**. **Desktop publishing software** (DTP) is used when more advanced layouts are required. Office applications and Desktop Publishing software is available for PC, Mac and mobile devices.

Standard office applications

Word processors

Word processors are used for making reports and newsletters. They are easy to use and focus on the writing with the addition of images or charts. Writing tools are available such as a spellchecker and grammar checker.

Spreadsheets

Spreadsheets allow the entry of numbers and text into cells. **Formulas**, that perform calculations, are entered into other cells. This allows calculations to be updated dynamically if any values change. **Charts** may also be created automatically from data on the spreadsheet.

Presentations

Presentations are a collection of slides that are played one after the other. Text, graphics, charts and infographics may be added. Objects may have an **animation** applied and slides may have **transitions**, such as a fade, applied to go from one slide to another.

Databases

Databases store information in a structured and organised way. Example uses include customer databases, school databases and product databases. **Forms** are used for **data entry**. **Queries** are used to search the data. **Reports**, which make use of queries, allow tables of information to be created and may include charts. These are often printed or sent as a PDF file.

Desktop Publishing (DTP)

Desktop publishing software is used for creating printed documents such as leaflets, newsletters, magazines or books. A PDF version of the documents may also be created.

Key features:

- Text and images are placed in frames.
- One **text frame** may be made to flow into another frame.
- Text is able to **flow** around images.
- There are advanced features for **typography** (the appearance of text).
- Rulers, grid lines and **snapping tools** all help to align objects on a page.
- Pages can be shown as **spreads** (two pages side by side) in the same way as they are in a book or magazine.

A lifeguard is currently running a training course in first aid. They would like to show key points and photos in slides on a projector screen in the training room.

(a) Name the item of software they would use to create the slides. [1]

(b) They will be creating training materials with further information. Explain suitable software to use. [2]

The lifeguard runs a small business training people. They want to store information about their students in a database.

(c) Give **one** database feature which will be useful when finding the email addresses of all people that went to a training course last week. [1]

(d) Describe how database forms could be used within the business. [2]

(a) *Presentation software.*[1]

(b) *A word processor*[1] *as it is easy to use*[1] *and allows text and images/photos/graphics to be added.*[1] *Desktop Publishing/DTP software*[1] *as it gives a high degree of control over the layout of the page*[1] *which allows a more professional document to be created.*[1]

(c) *Queries.*[1]

(d) *A database form would allow the data entry of new students*[1] *and show the details of any existing students.*[1] *It could also be used to add/edit/view details of any training sessions.*[1]

5.3

DIGITAL DEVICES

Digital devices have a variety of different characteristics which make them more or less suited to different scenarios. The most common digital devices are given below.

Smartphones

Smartphones connect to both Wi-Fi and 4G/5G mobile phone networks.

- ⊕ Highly portable devices that fit in a pocket.
- ⊕ Touchscreens are usually used.
- ⊕ Apps are easy to download.
- ⊖ Small screen size.
- ⊖ Not suitable for certain types of use, such as desktop publishing.

Tablets

Tablets connect using Wi-Fi. Some tablets allow 4G/5G connections.

- ⊕ Portable devices – smaller than laptops, larger than smartphones.
- ⊕ Touchscreens are larger than smartphones.
- ⊕ A wider range of apps than smartphones.
- ⊖ External input devices may be required for business use, such as keyboards or external displays.

Desktops and laptops

Desktop computers (**PCs** or **Macs**) commonly used in businesses and for professional work.

- ⊕ Very wide range of apps/software.
- ⊕ Large storage area and fast.
- ⊕ Multiple large screens may be used.
- ⊕ Easily upgraded.
- ⊖ Larger machines that aren't portable.
- ⊖ Require access to mains power.

Smart TV

Smart TVs are digital TVs that connect to the Internet.

- ⊕ Apps may be downloaded and installed.
- ⊕ User interface is easy to use.
- ⊕ Works with a range of streaming apps.
- ⊕ Web browsing may be possible.
- ⊖ Difficult to enter text with a remote control.

Smartboards

Smartboards are commonly used for presentations in trainings and classrooms.

- ⊕ A large screen allows viewing from a distance.
- ⊕ Teachers, trainers and learners can directly interact by touching the screen.
- ⊖ Expensive.
- ⊖ Require a large empty wall to install.

A delivery company needs to update all drivers with their routes whilst they are out delivering packages.

Explain one suitable digital device [2]

A smartphone[1] as it is portable[1] / can connect to a mobile network[1] / is easy to dock on a van dashboard.[1] A tablet[1] as it is relatively portable[1] / has a large screen to display the route.[1]

5.4.1

TYPES OF DISTRIBUTION CHANNEL

Once documents, media or content have been created they need to be shared with others. This takes place through **distribution channels**. Key distribution channels to be aware of are given on the next two pages.

Cloud

Documents such as reports and presentations often need to be shared with other people in an organisation, or with their business contacts and customers.

A good way to distribute these is using **cloud storage**. This allows the documents to be shared using a web link. The documents may also require the use of a password or a user to log in.

Using such a system is easy to use and can be accessed anywhere in the world. Fast Internet connections are required and the cloud storage provider must be trusted to store documents securely.

Email

Email is a common way to distribute documents and media. Text and graphics can be combined into email and attachments of other documents may also be added. Links to other media online may also be included.

Email can be **tracked** to see if people open the email or click a link. Most people check their email daily or have **notifications** which alert them when email arrives.. One issue with email is that they may end up in a spam folder.

Messaging

Messages are sent via text messages, social media, and **messaging** apps. These allow messages to be sent between customers, colleagues or the public. For instance, a delivery company may send a text message with the time a package may arrive. A company may use a messaging facility on their website for customers to obtain **live support** .

VoIP

Voice over Internet Protocol (VoIP) is used to deliver voice and video calls. This can be used for meetings where media such as video or a presentation may be distributed.
All participants will need a fast and reliable Internet connection.

OCR Cambridge Nationals J836 (R050) **IT — Topic 5**

Multimedia

Multimedia is the term used when two or more types of **media** are combined together. Examples of how multimedia can be distributed include:
- A website that allows customers to read a description of a product and also view photos and videos of it.
- A presentation which combines infographics, text and images. The presentation is then distributed by a website, cloud storage or email.

Media may be distributed through multiple channels. For example, a newsletter could be sent to customers via email. Customers could also download the newsletter from the company's website. A customer could also be sent the newsletter through a messaging app.

Websites

Websites are used to share many types of media. They can be accessed anywhere in the world by members of the public.

Organisations may make use of an internal website that is only available to employees inside their own network. This is known as an **intranet**.

Websites are typically very interactive as they have links for users to find different content. Menus, forms and the ability to launch and play video are additional ways that users often interact with websites.

An organisation is responsible for their own website design and content which means they have control of everything. However, it also means that if the website has a problem, the organisation will need to solve it themselves.

Mobile apps

Mobile apps include games, **productivity apps** (which help users become more productive at work), social media apps and fitness apps. These apps usually make use of multimedia and allow the distribution of content. For instance, a user of a fitness app may distribute new running routes to other users. By contrast, a business may distribute adverts or infographics through social media posts.

Mobile apps are convenient and engage users through the use of notifications. Whilst a company can develop their own app, there is a cost and not all customers will want download or pay for the app.

---

1. Describe how VoIP could be used to share a presentation. [2]
2. Give **two** benefits of using a website as a distribution channel. [2]
3. Describe how a delivery company could keep users updated with information about their package through the delivery network. [2]

   1. During a video conference call/meeting[1] the presenter could start showing their presentation[1] which would then appear on all the other screens.[1]
   2. A wide range of media may be delivered.[1] A photo of the delivery location may also be provided.[1] The website could be accessed anywhere in the world,[1] using many different devices.[1] The company is in full control of the website and its contents/appearance/functionality.[1]
   3. They could send text messages/update a website/send push notifications in an app[1] at key points in the journey[1] such as when the package is collected/arrives at the distribution centre/is out for delivery/is delivered.[1]

5.4.2

DISTRIBUTION CHANNEL CONNECTIVITY

The types of distribution channel given on **pages 49-50** all need an underlying method of **connectivity**. There are different characteristics, advantages and disadvantages for each connection method. The different characteristics affect which connection method is most suitable for a given scenario.

Wired

Computers and servers are usually connected to networks with **wired connections**.

Wired connections are cheap to run, fast and reliable. However, there is often a significant cost to installing wired networks and computers cannot be easily moved. Wired connections are not suitable for mobile devices.

Ethernet cables

The most common type of connection for home and business computers is a copper twisted pair **Ethernet cable**. Connections are made directly from a computer to a switch or hub up to 100m away.

Fibre optic cables

Fibre optic cables are commonly used to connect homes and businesses to the Internet. Fibre optic cables have extremely high **bandwidth** (the amount of data transmitted per second) and connections can be many miles apart.

Wireless

Wireless connections are easy to install as no cables need to be used. They usually make use of radio waves or microwaves to transmit data. They are particularly important for mobile devices to allow them to be portable. Issues with wireless connections include **signal quality** if moving or located far away from a **wireless transmitter**.

4G and 5G

4G and **5G** are technologies to connect mobile phones and tablets to **mobile phone networks** and the Internet. The G stands for generation. 4G networks are usually fast enough for web browsing and video, however they are not as fast as Ethernet and fibre optic connections.

5G connections are comparable to the speeds used in most homes and businesses for Ethernet and Fibre optic connections. This allows 5G to replace many wired home connections and also makes it easy to connect IoT (Internet of Things) devices.

Wireless continued

Wi-Fi

Wi-Fi is a technology commonly used to connect laptops, tablets and mobile phones to home and business networks. Connection speeds are comparable with 5G and Ethernet connections. Devices need to be relatively close to the Wi-Fi **wireless access point** (**WAP**). Signal strength is also affected by obstacles such as walls. The speed of a Wi-Fi network will reduce as more devices are added.

Wi-Fi access points are usually integrated into home routers. By contrast, businesses often use separate wireless access points then have a wired connection to their network and router.

Mobile Wi-Fi hotspots create a small network using Wi-Fi. They then connect to the Internet using 4G and 5G mobile phone networks. Hotspots may be powered, such as on a train, or be battery powered for portability. Many smartphones can be turned into mobile Wi-Fi hotspots allowing other phones and mobile devices to connect to the Internet through them.

Bluetooth

Bluetooth is a method of connecting devices that are under 10 metres away. It is used to connect mobile phones to wireless headphones, smart speakers and car entertainment systems. Bluetooth is also used for connections to smartwatches and fitness trackers.

Bluetooth connections don't require much power making them suitable for small portable devices. Whilst connecting Bluetooth devices is easy, it is not very secure. Data transfer speeds are also relatively slow.

1. Give **two** limitations of Bluetooth. [2]
2. An electrician needs to access the Internet on a laptop when at customers' houses. Describe a suitable method of connection for the laptop. [2]

   1. It has a very short range,[1] slow data transfer speeds[1] and a lack of security.[1]
   2. The laptop could be connected to a mobile hot spot[1] using Wi-Fi.[1] The mobile hot spot would then connect to the Internet using 4G or 5G.[1]

5.5

AUDIENCE DEMOGRAPHICS

The type of digital communication created and the way it is presented will depend on the **target audience**. This is the group of people that the communication is aimed at. Each target audience will have a range of characteristics that need to be considered.

Age

The age of users will significantly affect the design of digital communications.

- Young children may need simpler text with clear font and brighter colour schemes and graphics.
- Older adults may prefer larger quantities of text with more subtle use of colour.
- The type of digital communication is often affected by age, for example, a presentation or report may be more suited to adults whilst an infographic, leaflet or social media post may be more suited to younger people.

Gender

Different genders may respond differently to design styles, colour and content. In many situations a neutral style will be appropriate so that it appeals to all. However, in some cases the gender of the users needs to be carefully considered.

Accessibility

Accessibility of digital communications needs to be considered. Some examples include:

- having a large print version of a leaflet for those with sight difficulties
- using closed captions or subtitles with video for those with difficulties hearing
- using **alternative text** (**alt text**) with images so that blind people with screen readers have the image described.

Location

Websites may be customised to different countries. This may involve adapting text or cultural references.

An internal report on an intranet may contain more detailed information to one presented to the general public.

A mobile phone manufacturer has produced a video of all the features of a new mobile phone they are launching. The presentation will be viewed by the sales team.

Explain **two** alterations that may be needed when the video is adapted to show new features to the general public. [4]

(If not already present) the video will need closed captions/subtitles added[1] for members of the public with hearing difficulties.[1] The content may be less technical[1] as the public may not understand more complicated details.[1] The video may need to use different graphics/background music/narration[1] to make it more appealing to customers.[1]

Topic 5

EXAMINATION PRACTICE

Section A style questions

1. One type of demographic is age. Give **two** other examples of demographics. [2]

2. A mobile app gives information about caring for pets for adults aged 18 or over.
   A new version of the app is being created that is more suitable for children.
   Explain **two** ways the app may need to be altered for this new audience demographic. [4]

3. An infographic has been created for a charity showing various statistics about how much time people spend online each week.
   (a) Explain the term 'infographic'. [2]
   (b) The charity wants the infographic to be seen by a large number of people.
   Explain **two** suitable distribution channels that the charity could use. [4]

4. A local swimming club wants to advertise their club to members of the public. They want to create a leaflet.
   (a) Give **one** advantage and **one** disadvantage of using a leaflet to advertise the swimming club. [2]
   (b) The leaflet will be created using desktop publishing (DTP) software.
   Give **two** characteristics that make this software suitable for the task. [2]

5. Give **one** reason why social media may not be an appropriate communication channel to reach an older audience demographic. [1]

6. Explain **one** benefit of using a presentation as a type of digital communication for a younger audience. [2]

Section B style questions

Susie's Sweet Shop wants to advertise new products to their customers. Susie wants to use a loyalty scheme to reward customers who buy products. The shop sells a range of sweets and chocolate that are suited to a younger demographic.

The sweet shop will be creating a website to show the products sold and the new loyalty scheme.

7 The following shows a visualisation diagram of the webpage that will show the new loyalty scheme for the sweet shop.

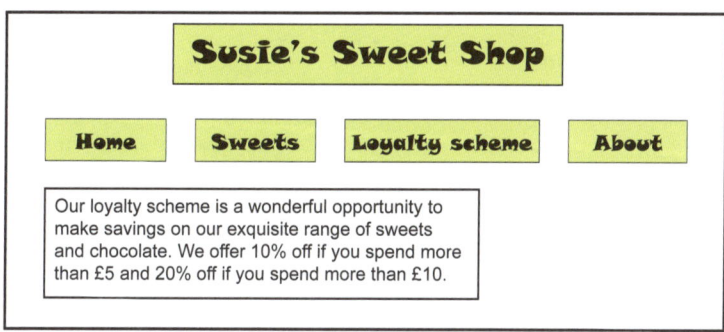

Identify **one** reason why this visualisation diagram may not be suitable for the demographic of the sweet shop and give a suitable improvement. [2]

8 Some customers will need to access the website on a smartphone in the shop.

One method of connectivity which customers may use to access the website is via 4G/5G.

(a) Give **one** disadvantage to children using 4G/5G to access the website. [1]

(b) Explain an alternative method of connectivity that Susie could provide for customers to access the Internet. [2]

9 Susie has an accountant who calculates her sales and tax to be paid.

(a) Give **one** type of generic software her accountant might use for these calculations. [1]

The accountant puts the final figures into a report along with charts and text explanations.

(b) Give **one** type of generic software her accountant might use for creating the report. [1]

(c) Explain a suitable distribution channel for the accountant to share the report with Susie. [2]

10 Susie will be introducing new products to her shop each month. A leaflet will be created which shows:

- The product name and price;
- An image and description of the product;
- A child's positive comment about the product.

Discuss the suitability of Susie using a leaflet to communicate this information.
In your answer, you must consider:

- The purpose of the leaflet;
- The advantages and disadvantages to Susie's Sweet Shop of using the leaflet. [9]

6.1

INTERNET OF EVERYTHING (IOE)

Traditionally, networks and the Internet have been accessed by laptops, desktops and mobile devices. Today, many other devices are connected to networks including fridges, cars, televisions and smart speakers. These devices, often called smart devices are referred to as the **Internet of Things** (**IoT**).

IoT devices are able to communicate with each other and automate tasks in a manner that is referred to as the **Internet of Everything** (**IoE**).

Technologies used in IoE

Although the Internet is a key technology used in the IoE, other technologies are commonly used to connect devices. These include:

- **NFC** (**Near Field Communication**) – used for contactless payments
- **RFID** (**Radio Frequency Identification**) – used for contactless payments, product tracking and animal tracking
- **Bluetooth** – used to connect smart speakers and wireless headphones
- **Zigbee** – a low cost, low-power wireless technology to connect home and building automation devices. Devices connect to a hub which then connects to the Internet.

Examples of IoE devices

Smart speakers/ smart assistants

Smart speakers allow music to be played and voice communication. They often contain smart assistants which are able to carry out basic commands such as adding to reminder lists, changing the volume of music or turning lighting on and off.

Smart appliances

Smart appliances include fridges, washing machines and Internet dishwashers. Once connected to the Internet, they can be accessed via an app on a smartphone.

Smart lighting

Smart lighting refers to light bulbs that may be turned on/off or have their colour changed. These typically connect to the network using Wi-Fi or Zigbee.

Smart watches

Smart watches connect to a smartphone or network using Bluetooth or Wi-Fi. They can make **contactless payments** via NFC, track health and send messages.

The Internet vs the World Wide Web

The **Internet** is the worldwide network that connects most computers and mobile devices. It is the infrastructure of cables, switches and routers that make up the Internet and allow different devices to transmit data between each other.

The **World Wide Web** (**WWW**) is one service that makes use of the Internet. The world wide web consists of web servers which transmit web pages, and web browsers which request and receive the web pages.

The four pillars of the IoE

The four pillars of IoE are each of the key areas that make up the IoE system.

Data is any values which are being reported back from devices. This ranges from a temperature reading to security camera footage.

People produce data, such as by pressing a button on an app. They also consume data, such as by watching a video or listening to a music track.

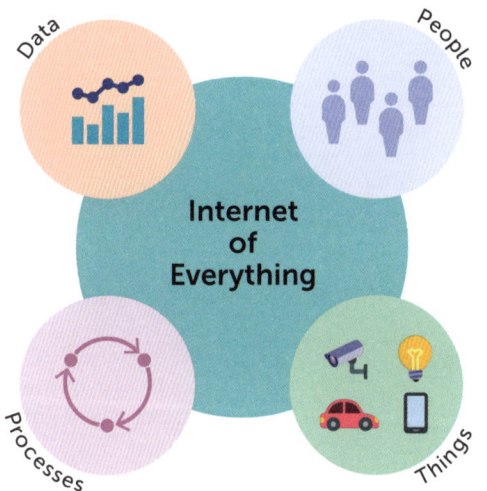

Processes are **automations** which are followed. For instance, a car getting near to a house may trigger home lights and heating to turn on.

Things are all the different devices that are part of the IoE. They include computers, smartphones, tablets and smart appliances.

1. One pillar of IoE is 'Things'. State the **three** other pillars of IoE. [4]
2. Computers, smartphones and tablets are all Things within the IoE. Give **three** other examples of Things. [3]
   1. People,[1] Processes,[1] Data.[1]
   2. Smartbulbs,[1] CCTV cameras,[1] cars,[1] smart heating systems,[1] smart appliances[1] (such as refrigerators or ovens). There are many more devices that could be given here.

6.1

INTERACTIVITY BETWEEN THE FOUR PILLARS OF IoE

Each of the different pillars are able to interact with one another.

An example of how people, things and data could interact in a process to prepare a house ready for an owner's arrival could be as follows:

Pillar	Interactivity between the pillars
Things	A car sensor detects that the user's car is 10 miles from home.
Data	Data is sent via a 4G/5G network to a server for processing.
Processes	A process is followed to ask the user if they want the heating turned on.
Data	Data is sent via a 4G/5G network to the user's smartphone.
Things	The smartphone asks the user if they would like the heating turned on.
People	The user says, "Yes please".
Things	The smartphone interprets the audio and sends the data.
Data	Data is sent to the smart home hub that it needs to turn on.
Things	The heating system is turned on.
People	The user experiences a warm home on arrival.

↓ Flow of interactions between the four pillars

Use of the World Wide Web and Internet in the IoE

Connected Things will make use of the Internet to send and receive data. They send data to either servers on the Internet or hubs located in homes, offices or organisations. Alternatively, they may send data to other devices on the IoE.

The World Wide Web is used to display web pages and content inside apps. The user interacts with their web browser or app which makes use of the World Wide Web. This data is then processed by the web server, which then sends data to other servers or Things.

A company makes use of RFID cards to keep a log of which employees are in the offices. It wants to use this data to automatically turn all the lights off when the last employee leaves.

Describe the interactivity between the four pillars of IoE that will occur to make this possible. [4]

When the RFID card is placed on the door/barrier[1] (People pillar), the door/barrier RFID reader[1] (Things pillar) sends data to a server[1] (Data pillar). The server detects that no one is left in the building[1] (Process pillar) and sends data to the hub in the offices[1] (Data pillar). The hub runs the process for turning the lights off[1] (Processes pillar) and sends data to all the smart bulbs to turn off[1] (Data pillar). The smart bulbs all turn off[1] (Things pillar).

IoE digital interactivity

Device to device

In the IoE, devices are able to communicate with other devices. This is known as **Device to Device** (**D2D**) or **Machine to Machine** (**M2M**).

For example, a smartphone and smart speaker may communicate with each other using Bluetooth.

Human to device

Humans and devices are also able to communicate with each other. This is known as **Human to Device** (**H2D**) or **Person to Device** (**P2D**).

For example, a person may give a voice command to a smart assistant on a smart speaker.

Tailoring devices to the needs of the user

Simple **automations** such as turning lights on at night and off in the morning were possible well before the IoE. The IoE allows far more sophistication and customisation in how devices and processes work.

For example, a user could configure their home lighting and heating with the following rules or processes that need following:

- Lights should turn on 30 minutes before it is dark.
- Lights should turn off at 11pm.
- At the weekend, lights should turn off at midnight.
- Home heating should switch on if anyone is in the house and the temperature is less than 18°C

The IoE system will now make use of data such as the current time, what time it gets dark in the current location and the current date, the temperature data from a smart home heating system and the location data given by smartphones that each home user carries with them.

Advantages and disadvantages of IoE

➕ Advantages	➖ Disadvantages
• Automation helps increase efficiency. • Repetitive tasks are automated and may be more reliably carried out which may improve personal wellbeing. • May help to keep people safer. • Time and money may be saved. • Most IoE devices and systems work with the Internet, which most organisations and people already have access to. • Supports a high-tech job market.	• Needs the Internet to work. • Those with lower digital skills may struggle to use IoE effectively. • Concerns over data security and hacking. • Impacts of cyber-security threats could be significant. • People become too dependent on technology. • Lower skilled jobs may be at risk of being replaced.

6.2

APPLICATIONS OF IoE IN EVERYDAY LIFE

IoE is used in a range of areas of everyday life. The following are those that need to be known for the exam. Remember that many devices in IoE are called 'smart', such as smartphones, smartwatches, smart meters. A huge number of other devices and processes also exist.

Energy management

Households and other organisations are able to use **smart meters** to improve their energy usage. Traditional meters have a display that must be manually read by a company or user. Smart meters are able to take frequent readings, usually hourly. These are then automatically reported back to the electric or gas company.

A small display in a house is able to show users their current, daily and weekly usage. These allow users to monitor their usage. Some homes and businesses may be charged different amounts for energy used at different times of the day.

➕ Advantages	➖ Disadvantages
• Users no longer need to take meter readings and send them to the energy providers. • Users are able to constantly monitor usage which helps to reduce energy consumption and costs. • Reduced energy consumption has environmental benefits.	• It needs a network connection. • There is a cost to upgrading old meters. • There is a risk of data theft and hacking. • There are privacy issues as more personal data is given with regular hourly readings rather than monthly ones.

Security

Newer smart meters take regular readings and also turn a meter on or off. If a hacker were able to alter a reading, then they could increase or decrease an energy bill. Alternatively, they could turn the electricity off or view personal data on energy usage. A cyber attack on smart meters could lead to disruptions and blackouts across the country. As the potential damage could be very large, smart meters make use of **authentication** and **encryption** when sending and receiving data. A lot of consideration is made to prevent security vulnerabilities.

Health

There are increasing ways in which IoE is used to monitor and improve people's health. Uses include:

- Automated monitoring of patients in hospital.
- Personal smart watches that track the user's heart rate, oxygen saturation and blood pressure during everyday tasks or exercise.
- **Smart scales** that communicate with apps to record body weight and percentages of water, fat and muscle.
- Automated detection and calling of emergency services following an accident.
- Monitoring health conditions such as diabetes or asthma and detecting emerging problems.
- Reminders to take medication and automatic ordering when more is needed.

➕ Advantages	➖ Disadvantages
• Improves quality of life for people with medical concerns. • Saves time in hospital meaning doctors and nurses can spend more time with patients. • More accurate and continuous measuring of medical conditions. • Direct feedback to medical professionals for faster diagnosis. • Data can be shared across the Internet allowing specialists to view data from anywhere in the world.	• Highly sensitive medical data will be transmitted, stored and processed. • Errors in a system may lead to the wrong diagnosis. • Malfunctions with equipment could cause harm. For example, requiring a person who has diabetes to inject insulin when not needed could have dangerous effects. • Some users may find the technology hard to use. • The technology may be expensive. • Some users may miss out on interaction with real people. For instance, elderly patients may have a weekly visit with a nurse replaced by using a machine.

1. Describe how IoE may help people to monitor and improve their health. [4]

    1. People (such as diabetics) could take blood tests which monitor blood sugar.[1] The results could automatically be recorded/sent to a medical professional.[1] If an issue is detected then medical support could be requested/given.[1] The history of readings could also be given to a doctor/nurse[1] leading to better diagnosis/treatment.[1]

        People could wear a smartwatch which detects blood pressure/oxygen saturation/heart rate.[1] This could monitor how these are effected during exercise.[1] It could help the user monitor their progress[1] and set goals to achieve.[1] If any concerns were detected[1] it could advise/automatically call for medical assistance.[1]

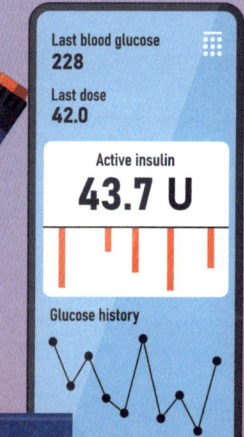

OCR Cambridge Nationals J836 (R050) **IT – Topic 6**

Manufacturing

Manufacturing is increasingly using the IoE in many different ways.

Production process efficiencies

- Robots and machinery are connected to the IoE.
- Each robot or machine is able to receive commands or report back precisely what it is doing.
- This allows a server to decide the most efficient way to manufacture products based on **real-time** data of what is happening on the factory floor.

Maintenance prediction

- Temperatures, pressures, sound and vibrations can be monitored to see if motors and other moving parts are operating normally.
- Minor issues can be detected and new parts ordered or **maintenance** scheduled before the part fails.

Warehouse management

- Products in a warehouse need to be **picked** and **packed**.
- Computers work out the most efficient route for robots or people to take to collect products.
- This enables the correct packages to be ready at the right time for each delivery lorry.

Tracking parts and products

- Parts can be tracked as they move through the supply chain. RFID tags and QR codes are often used in tracking.
- Finished products are **tracked** from the warehouse to the customer.

⊕ Advantages

- Robots and machinery can be remotely monitored and controlled.
- Reduction in downtime as parts can be ordered before they fail.
- As all IoE connected machines and robots communicate with each other, improvements can be made in the production process.
- Improvement in efficiency.
- Quality improvements are possible.
- On-demand production and ordering of stock.
- Reduction in storage costs of unnecessary products.
- Reduced waste from manufacturing.
- Energy efficiencies.

⊖ Disadvantages

- Reduced need for people may mean job losses.
- Robots and machinery that connects to IoE may be very expensive to buy.
- Workforce will need retraining.
- Hacked systems could lead to significant disruption and risks to safety.

---

2. Describe how IoE could be used in a warehouse for an online supermarket. [4]

> 2. Products could be tracked as they come into the warehouse[1] and the total number of products stored in a database.[1] When an order is received from a customer, the computer system will determine the most efficient order to retrieve the products[1] and then send instructions to robots/humans to pick the products.[1] Robots can feedback their exact location[1] so that further optimisations can be made.[1] Orders are ready in time for the correct lorry/van to depart.[1]

Military and emergency services

Military uses

The **military** includes the **army**, **navy** and **air force**.

There are many uses for IoE within the military. Three examples are **drones**, **robots** and **head-up displays** (**HUD**).

Military drones include **unmanned aerial vehicles** (**UAV**) which are used for surveillance. A remote operator is able to operate the vehicle remotely from another country.

Head-up displays may be used in soldier headsets allowing **Augmented Reality** (**AR**) information such as locations and distances to be projected in the view of the headset.

Robots are able to remotely detect and deactivate landmines.

Emergency services

Emergency services are able to make heavy use of IoE to improve patient care, health and responses. For instance, eCall SOS is a safety feature fitted to most cars and vans. If a vehicle is in an accident that triggers airbags, the location of the vehicle is automatically sent to emergency services. This allows police and ambulance services to be dispatched immediately even if the driver is unconscious. It also prevents emergency services wasting time going to an incorrect location.

Coastguard	Carries out search and rescue at sea. Helps enforce maritime law and security of the sea.
Fire service	Aim to protect life and property in the event of fire. Helps rescue from traffic accidents and other emergencies.
Ambulance	Responds to medical emergencies.
Mountain rescue	Recovers climbers and walkers injured or lost on hills or mountains.
Police	Aim to protect life and property and maintain law and order.

➕ Advantages

- Improved safety.
- More accurate locations help emergency services locate people faster.
- Real-time data helps to make better decisions and improve efficiency.

➖ Disadvantages

- False alarms could lead to unnecessary call outs for emergency services.
- If systems fail there could be life-threatening consequences.
- A breach in security could result in operational downtime, financial loss, loss of life and reputational damage.

Smart devices

Smart devices are any devices wirelessly connected to the Internet or networks. The key feature of a smart device is that, unlike a traditional digital device, it communicates with the IoE.

Home

Home smart devices include:
- Smart video doorbells.
- Smart alarms.
- Smart lighting.
- Smart robotic vacuum cleaners and lawnmowers.
- Smart speakers and virtual assistants.
- Smart appliances (such as ovens and refrigerators).
- Smart thermostats (for home heating).

Business

Business devices are often very similar to home devices. For instance, smart alarms and smart lighting may be used. The following additional items are also common:
- Smart locks.
- Credit and debit card readers.
- Smartphones and tablets.

Businesses often have more reliable devices and services than consumers as any loss of connection may create significant losses for the business.

Personal

Personal smart devices include:
- Smartphones.
- Smart watches.
- Personal health and fitness devices.
- Wireless headphones.

➕ Advantages

- More efficiency.
- Increased security.
- Automated stock orders.
- An ability to control many smart devices from one smartphone.

➖ Disadvantages

- Data is sensitive and hacks can leave people or businesses vulnerable.
- There is a cost to installing and maintaining equipment.
- Some people don't like smart devices, especially where human contact is lost.
- If the technology fails, there can be significant costs to a business or inconvenience for users.

Transport

There is a heavy use of IoE in transport areas today.

- **ANPR** (**Automatic Number Plate Recognition**) is used to read car number plates. This enables fines to automatically be issued where car parking hasn't been paid. The same technology is used to open car barriers when tickets have been bought online.
- Cameras are able to monitor the amount of traffic on motorways. If congestion is seen, signs can automatically be altered to slow traffic down or close roads entering the motorway.
- Navigation apps for cars may send data about their location and speed to a central server. It is then possible to work out where congested roads are so that other drivers are able to avoid these areas.
- Planes have **engine health management** (**EHM**) systems which are able to find problems. When an issue is detected, hundreds of hours of information could be sent back to the manufacturer for analysis.
- Data from vehicle **engine control unit** (**ECU**) can be reported back to the manufacturers to help improve efficiency and safety.
- Train stations make use of RFID to let people enter or exit ticket barriers. A phone or RFID card is used to deduct the correct amount for the journey from the passenger.

➕ Advantages	➖ Disadvantages
- More efficient parking for drivers. - Parking infractions are more efficiently detected. - Incorrect use of bus lanes is reduced allowing faster journeys for public transport users. - Safer transport through reporting of data back to manufacturers. - More efficient travel networks.	- Software issues may lead to inappropriate decisions being made. - The infrastructure costs money to install. - An IT or Internet failure could lead to catastrophic accidents. - System security is important as significant amounts of personal data will be transmitted.

3. Describe how the IoE helps to provide efficient car navigation. [2]

3. Cars communicate their current location and speed to a central server/IoE service.[1] As many cars report this data, the service is able to find any sections of a route which are congested[1] and send this data to the satnav/smartphone/in car navigation system.[1] If necessary, new, faster routes are recalculated.[1]

Topic 6

EXAMINATION PRACTICE

Section A style questions

1. There are four pillars of the Internet of Everything (IoE). One pillar is named Things or Devices. Name **two** of the other pillars. [2]

2. IoE digital interactivity occurs as device to device (D2D) and human to device (H2D).
   A user makes use of their smartphone to control their home lighting.
   (a) Describe **one** digital interactivity which occurs from device to device. [1]
   (b) Describe **one** digital interactivity which occurs from human to device. [1]

3. Identify **one** drawback of the Internet of Everything. [1]

4. Give **two** examples where the IoE may be used in a gym. [2]

Section B style questions

A delivery company decides that they want to install a new 'smart office' at their head office.

As part of the smart office, they want to provide a better working environment for employees by installing an automated heating system.

5. Identify **two** smart devices that may be used in the office as part of the heating system and describe how each one may be used within the system. [4]

6. The office will make use of CCTV cameras which connect to the IoE.
   (a) Give **two** advantages of connecting these cameras to the IoE. [2]
   (b) Give **two** disadvantages of connecting these cameras to the IoE. [2]

The delivery company has approximately 1000 drivers making deliveries each day.

7. Sometimes delivery drivers are involved in accidents.
   Describe how the IoE could be used to improve the response time for an ambulance or the police to arrive. [4]

8. Some delivery drivers travel long distances before making a stop.
   Describe how an in car entertainment system could make use of the IoE to make the journey more pleasant for the driver. [4]

9. Many customers would like to have up to date tracking information about products as they are being transported and delivered.
   Discuss how the Internet of Everything (IoE) could be used when tracking products.
   In your answer, you must consider:
   - How different devices (IoE Things) could be used;
   - The advantages and disadvantages of using the IoE for tracking products. [9]

EXAMINATION PRACTICE ANSWERS

Topic area 1: Types of design tools

1.

Flow chart symbol	Meaning
parallelogram	Input/output symbol.[1]
diamond	Decision symbol.[1]
rounded rectangle	Start/stop symbol / Terminator.[1]
rectangle	Process box/symbol.[1]

[4]

2. **C** One or more arrows indicating direction of flow.[1] [1]
3. Library mind map,[1] presentation mind map.[1] [1]
4. Branches/linking lines/arrows,[1] nodes,[1] sub-nodes,[1] keywords,[1] images,[1] colour,[1] shapes/bubbles.[1] [2]
5. (a) It looks like the finished product[1] so feedback can be given before the product is designed.[1]
   Annotations are included[1] to give more information / help justify design choices.[1]
   (b) Advantages of wireframes (compared to visualisation diagrams):
   Wireframes are quick to create[1] as they are usually black and white / can be made with pen/paper / use simple symbols[1] this saves the designer time[1] and allows for quick alterations.[1]
   Disadvantages of wireframes (compared to visualisation diagrams):
   Visualisation diagrams have more detail / colour / detailed sketches / annotation[1] which helps people get a better idea of the finished product[1] and also helps a designer/developer to understand exactly what they will need to create.[1]

[4]

(c) Use the marking grid on **page 76** to help with marking this question.
Up to a maximum of 4 marks for layout and a maximum of 4 marks for content. [8]

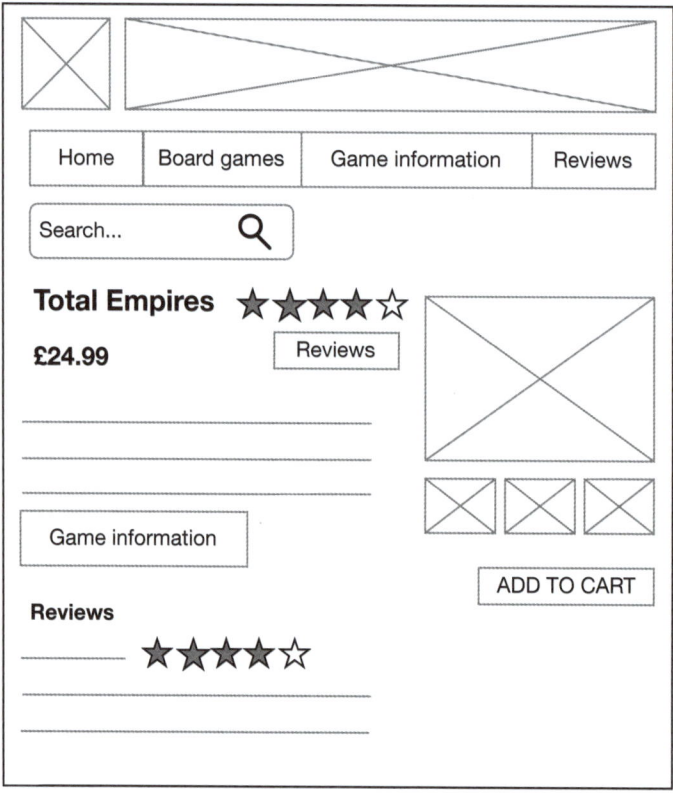

Indicative content

Layout:
- Screen for one product.
- Header / banner image.
- Image placed for a logo.
- Horizontal menu towards the top of the page, and/or vertical menu to the left of the page.
- Clear sections for content.
- Clear product title.
- Well laid out images that align with other objects.

Content:
- Menu / links to board games, game information, reviews.
- Buttons / links for navigation.
- Ability to buy or add to cart.
- Image for logo.
- Image for product.
- Additional product images.
- Title of product.
- Rating of product.
- Lines to show text for product description and reviews.

6. Use the marking grid on **page 76** to help with marking this question.
   Up to a maximum of 4 marks for layout and a maximum of 4 marks for content. [8]

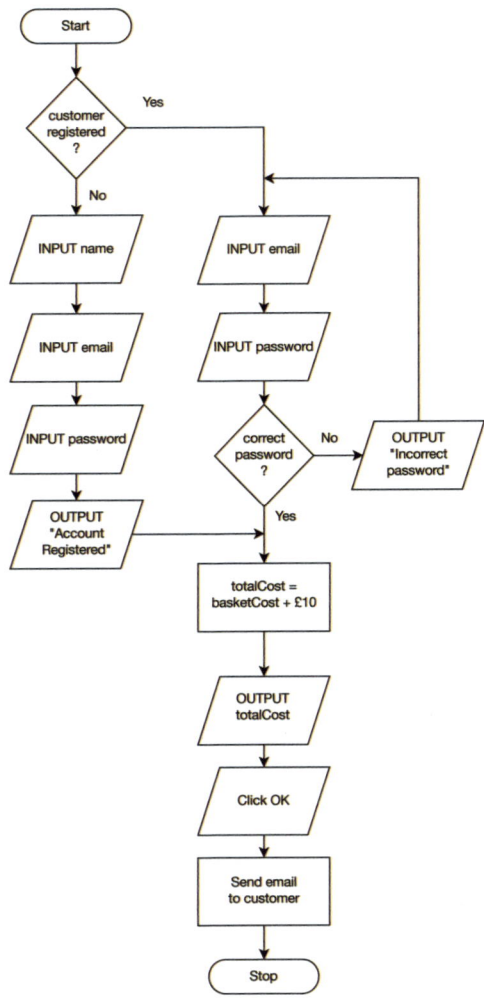

Indicative content
Layout:
- The flow chart has a clear start.
- The flow chart terminates/ends/stops.
- Symbols have a logical layout with different parts of the problem in different areas and a top to bottom flow.
- Arrows are used to show direction of flow.
- Correct labels are used for any decisions.
- Loop if an incorrect password is entered.
- Whether the user is registered or a new customer, both go to the same part of the flow chart for calculating the total cost and sending the email.

Content:
- Correct symbols are used for input, output, decision, process and terminators (start/stop).
- Inputs for name, email and password if the user is not registered.
- A suitable output to let the user know the account has been registered.
- Inputs for email and password if the user is registered.
- Check the password in a decision box.
- A suitable error message output if the password isn't correct.
- A process box to add the £10 delivery cost to the basket cost.
- Output of the total cost.
- Input symbol for clicking the OK button.
- A process box for sending an email to the customer.

Topic area 2: Human Computer Interface (HCI) in everyday life

1. Mouse,[1] keyboard,[1] voice.[1] [2]

2. Most methods of interaction (such as touch, gestures, keyboards, mice) involve the use of hands/arms/physical gestures.[1] If someone has physical disabilities, they may struggle to use these methods of interaction but be able to speak perfectly well.[1] Hence using voice as an HCI method, such as for voice commands/dictation, will allow them to easily make use of the HCI.[1] [2]

3. New users will find it easier to learn how to use/navigate.[1] This will lead to fewer problems/issues when using it[1] which will result in reduced stress for users / less need for customer support / higher satisfaction.[1] [2]

4. Command line interface / CLI.[1] [1]

5. (a) The use of a GUI will require more processing power/a faster processor[1] to run the additional programs/instructions/code that controls the graphics[1] and more memory/RAM[1] to store the graphics displaying on the screen.[1] More memory/storage space[1] will also be needed to store the extra programs/instructions/code that generate the graphics.[1] [4]

   (b) An LCD display[1] is inexpensive[1] and commonly used in phones.[1]
   An OLED display[1] will look bright/attractive[1] and is commonly used in phones.[1]
   An electronic paper screen[1] will use very little power/allow the battery to last for a long time[1] / allow the screen to be easily seen in daylight.[1] [2]

   (c) **Advantages for making phone calls:**
   A swipe gesture would allow children to easily/intuitively scroll through an address book.[1] Children would then be able to select the number they wish to call by pressing the name.[1]
   An onscreen keypad could be used to show the numbers 0-9.[1] These could then be pressed to dial a number.[1]
   If a finger is used, this has the advantage that it won't be lost/doesn't require a stylus to be carried around.[1]
   Advantages for writing texts:
   A virtual keyboard can be shown[1] allowing the user to enter the message with their finger.[1]
   As the user types, a set of predicted words could be shown[1] allowing the user to select the correct word they intended[1] / will make entering the text faster/more accurate.[1] [4]

   (d) A mouse requires a flat surface/mousepad to work[1] / is a bulky device.[1] A cable/external socket is required.[1] [2]

   (e) Linux / Android[1] as it is a free operating system[1] / as it is open source which allows it to be customised to the developer's requirements.[1] [2]
   [**Note:** Allow iOS only if the answer shows that it is proprietary and therefore the phone developer would need permission/need to develop with the owner – Apple.
   Ubuntu is a desktop operating system. Ubuntu Touch would be a possible operating system for a mobile device.
   Windows is a desktop operating system that is also suitable for tablet devices and servers. Windows Phone/Windows 10 Mobile was an operating system used for mobile phones but was discontinued in 2017.]

Topic Area 3: Data and testing

1. Data is raw facts.[1] Information is the data plus meaning/structure/context.[1] [2]

2. (a) Validation tools help to make sure that data meets certain requirements/rules/is sensible/reasonable.[1] [1]

   (b) The same data is entered twice.[1] The data is then compared[1] and if they match they are valid[1] and if they don't match the data is invalid.[1] [3]

   (c) Any 3 from: [3]
   - Data type check
   - Format check
   - Input mask
   - Length check
   - Drop down list
   - Radio buttons
   - Tick list
   - Lookup
   - Presence check

3. (a) **Email**[1]: They could email local residents asking for their opinions the beaches.[1] This would allow them to get direct responses from a range of local residents.[1] The email could contain a link to an online form.[1]
Interview[1]: An interviewer could stand on the beach and ask people their thoughts.[1] This would capture the feelings of people on that day on the beach.[1]
Questionnaire/survey[1]: They can post a questionnaire to local residents asking specific questions about the beaches.[1] This would capture the views of local people.[1] The survey would then need to be returned or collected.[1]
Online questionnaire[1]: Public can complete a form on a website[1] which, when submitted will automatically store the data in a database/spreadsheet.[1] People may be directed to the website through an email/letter/newspaper/social media post.[1] [3]

(b) **Books/magazines/newspapers**[1]: The local council could read books/magazines/newspapers to find out any reported views on the beaches[1] along with comments from local residents/tourists.[1]
Government statistics[1]: The council could look to find statistical data on public satisfaction of the beaches.[1] Other government statistics could be useful such as water cleanliness.[1]
Websites[1]: The local council could go to websites such as social media/review sites/tourist sites[1] to see what the general public is saying about the beaches.[1] [3]

4. (a) They will be able to share data with others (using hyperlinks to the location).[1] The data will be backed up by the cloud storage provider.[1] The data will be held securely by the cloud storage provider.[1] They will be able to access the data from anywhere in the world (where they have access to an Internet connection.[1] If they lose a device, they will still have all the data stored on the cloud.[1] [2]

(b) If it is dropped then it may break.[1] There are moving parts (the disk platter and drive head). If these fail the data will be lost.[1] The device could easily be lost[1] or stolen.[1] A cable is required to connect the drive to the computer.[1] [2]

(c) Network-attached storage (NAS),[1] portable USB flash drive.[1] Portable solid state drive.[1] [1]

5. (a) A test user could be given a series of problems/tests to complete[1] and then watched/recorded by a developer/tester to see how well they navigate/use the website.[1] [2]

(b) Integer,[1] decimal/real.[1] [1]

(c) Extreme test data will test valid data at the boundary of the ages required. In this case: 1,[1] 15.[1] [2]

Topic area 4: Cyber-security and legislation

1. (a) The hard disk drive/solid state drive/computer's storage[1] will be encrypted[1] by the malware. The user then won't be able to access the data on the drive[1] due to it being encrypted.[1] There will be a request for payment[1] (usually by bitcoin or a cryptocurrency). If the ransom is paid, the data will be decrypted.[1] It is possible that the money will be taken, but the attacker won't decrypt the data.[1] [3]

(b) Worm.[1] [1]

(c) Adware,[1] botnet,[1] spyware,[1] Trojan Horse,[1] virus.[1] [2]

2. White hat hacking is where a hacker tries to find vulnerabilities in computer systems[1] and has permission to do this from the owner.[1] The hacker will identify weaknesses[1] so that the company can make their systems more secure.[1] [2]

3. Denial of Service/DoS attack.[1] [1]

4. (a) Currently a password is used which uses just one factor[1] (something the user knows). A different factor would be something the user has / something the user is.[1] By using an additional/second factor, such as fingerprint/facial recognition (something the user is)[1] / a card reader / confirmation code by mobile text/email (something the user has)[1] security would be increased. [2]

(b) Biometric devices,[1] fingerprint scanner,[1] retina scanner,[1] facial recognition,[1] keypads,[1] RFID.[1] [2]

(c) Access rights and permissions give the ability to restrict the files/applications/software that can be run by each employee.[1] For instance, the software (or software feature) that allows the transfer of money could be restricted to just senior staff.[1] Junior staff would only be given the rights/permissions to view account information so that they can respond to queries.[1] [2]

(d) The original information is encoded[1] in such a way that it cannot be .understood (if intercepted) by a hacker as it is transferred across the Internet.[1] Once it arrives, a password/key is used to decrypt the data[1] so that it can be understood again. [2]

(e) The physical firewall is installed between the Internet connection and router for the internal network.[1] It checks packets of data to see if they are harmful.[1] If a harmful packet is found, it will be discarded/dropped so that it cannot harm any computers on the internal network.[1] [2]

(f) The Data Protection Act.[1] [1]

Topic area 5: Digital communications

1. Accessibility,[1] gender,[1] location.[1] [2]

2. The images/text/buttons may need to be larger[1] to make them more appealing/easier to read/easier to press.[1]
   The text may need to use simpler language[1] so that it can be more easily understood by children.[1]
   Complicated concepts/instructions may need to be removed/further explained[1] so that younger children can understand them/not be intimidated by them.[1]
   The design/colours may need to be bolder/brighter[1] to attract attention/be more appealing.[1] [4]

3. (a) An infographic combines images/icons[1] with text and information[1] in a way that makes it more interesting[1] and easier to understand/remember.[1] [2]
   (b) The infographic could be emailed to a mailing list of interested people/charity members.[1] These people would be receptive to the content[1] and by using email it would be likely they would read it/be notified about it.[1]
   A message could be posted on social media/a messaging app[1] which could then be reposted further.[1]
   The infographic could be uploaded to the charity's website.[1] This would allow anyone visiting the site to easily download it.[1]
   Other distributions may be acceptable if they have a suitable explanation that links to the scenario. [4]

4. (a) **Advantages:**
   - A leaflet is easier to create[1] (compared to other types of digital communication such as a website).
   - The leaflet can be printed (and therefore is easy for the public to take).[1]
   - A printed leaflet would be accessible to people who don't have a computer/Internet.[1]
   - The swimming club will have full control over the layout/house style of the leaflet[1] (unlike if they used a 3rd party system such as social media).
   - A template could be used to make the leaflet appear more professional.[1]

   Disadvantages:
   - There is a cost to printing leaflets[1] (especially if professionally printed).
   - The leaflet may become out of date quickly[1] (especially if it has prices).
   - If it is thrown away, the user won't be able to find the information[1] (which they could if it was on a website).
   - It isn't possible to track how many people read/interact with a leaflet[1] (whereas it is possible with social media posts or websites). [2]

   (b) This allows images/text to be added as frames.[1] Text can be made to flow from one frame to another.[1] There are precise positions of the layout[1] and advanced tools for adjusting the typography.[1] Guides allow objects/frames to be accurately positioned.[1] Editing in spreads (2 pages side by side) is possible.[1] Master pages are available[1] which help to keep pages consistent/make production more efficient.[1] [2]

5. Some older people may not use social media[1] or may not have digital devices needed to access it.[1] [1]

6. Multimedia/sound/video can all be added[1] which helps to add interest/improve engagement/improve understanding of topics.[1] Animations/transitions[1] help to add interest/engage viewers.[1] [2]

7. The web page is text heavy[1] and would benefit from images / bullet points for key points.[1]
   The text content is more appropriate for an older audience[1] and should be written using simpler language.[1]
   The design is not appealing due to lack of colour/images[1] and would benefit from the inclusion of photos of sweets/a graphic for the 20% reduction/bright graphic shapes.[1] [2]

8. (a) The signal may be unreliable.[1] There may be an additional cost to using 4G/5G.[1] Some children may not have access to mobile broadband.[1] [1]
   (b) Setting up a Wi-Fi[1] in the shop would allow children to connect to the wireless access point/WAP[1] which would be convenient/have no cost for the children.[1] [2]

9. (a) A spreadsheet. [1]
   (b) A word-processor.[1] (Allow Desktop Publishing/DTP). [1]
   (c) Cloud storage[1] would allow the report to be shared with password protection.[1]
   The report could be attached to an email[1] which would only be seen by Susie / would be easy to access.[1]
   The report could be shared via a secure area of the accountant's website[1] which Susie would log in to access.[1]
   [**Note:** *A website is, in general, not suitable for sharing confidential information. This answer would be acceptable as long as it has been made clear that a log in is required therefore keeping the information confidential.*] [2]

10. The following shows examples that could be given for the purpose, advantages and disadvantages. The question will be marked using the levels based marking criteria at the bottom of the page.

Purpose:
- To inform customers/children about new products and their prices.
- To encourage customers/children to come to the shop/buy products at the shop.
- To give positive feature about each product from a customer's/child's point of view.

Advantages:
- Susie's logo, and shop name can be included in a prominent position on the leaflet's first page.
- Text, photos and graphics/illustrations can be combined and made to appeal to the target demographic/children.
- A template can be used so that all pages in the leaflet are consistent/look professional.
- The (printed) leaflet can be viewed without the need for devices/Internet.
- Directions to/address of the shop can be given.
- The leaflet could be left in other shops or locations that children are likely to go to.

Credit will be given for any other suitable response.

Disadvantages:
- Product information/prices may change but cannot be updated on the leaflet without re-printing more leaflets.
- If there is no online version of the leaflet, customers may lose it/not take it, then be unable to access the information.
- If a template isn't used then each page may be inconsistent.
- It is expensive to print leaflets.
- If the leaflet is shared as a digital file, it may be hard for customers to read (for example, a 3-fold leaflet may have sides out of order).
- A printed leaflet can only be shared with one person at a time (whereas a digital file can be shared with many people at once).
- A digital version of the leaflet will need a device to read it and an internet connection to download it.

Credit will be given for any other suitable response.

Level 3 (high) 7-9 marks
A thorough discussion which shows detailed understanding:
- Detailed knowledge and understanding of the purpose of the leaflet.
- More than one advantage and disadvantage to the shop of using the leaflet for advertising are explained.
- Relevant and appropriate examples are given.
- Consistently used appropriate terminology.

Level (2) (mid) 4-6 marks
An adequate discussion which shows sound understanding:
- Sound knowledge and understanding of the purpose of the leaflet.
- At least one advantage and/or one disadvantage to the shop of using the leaflet for advertising are described.
- Some relevant examples are provided although these may not always be appropriate.
- Some use of appropriate terminology.

Level (1) (low): 1-3 marks
A brief discussion which shows limited understanding:
- Limited knowledge and understanding of the purpose of a leaflet.
- Few advantage(s) and/or disadvantage(s) of a leaflet in general are identified.
- Little or no use of appropriate terminology.

0 marks
No answer worthy of credit.

Topic area 6: Internet of Everything (IoE)

1. Data,[1] Processes,[1] People/Humans.[1] [2]

2. (a) Device to device: The smartphone communicates with the home lighting / a smart bulb / a server on the Internet that controls the light.[1]
   (b) Digital interactivity: A person touching the smartphone/selecting a button on the smartphone.[1] [2]

3. If the Internet/network connection fails then the IoE will fail to work correctly.[1] Some people may struggle to use the devices/IoE effectively.[1] There is a data security/hacking risk.[1] Cyber-security threats will be significant.[1] People may become too dependent on the technology.[1] Lower skilled jobs may be at risk.[1] [1]

4. A person uses a smartwatch/fitness band / a fitness machine[1] which monitors their heartbeat/blood pressure/oxygen saturation.[1] Smart weighing scales[1] are able to monitor/track weight/muscle mass/bone mass/water concentration.[1]
   Door entry systems/barriers can be opened[1] using a smartcard/PIN.[1]
   A trainer could record progress[1] on a tablet/smartphone.[1] [2]

5. Smart plugs.[1] A heater could be connected to the plug and turned on/off remotely/at night.[1]
   A smart thermostat[1] which sends data to a heating system hub[1] which turns the heating on/off when it is too hot/cold.[1]
   A heating system control unit[1] which allows the user to set the desired temperatures/times.[1]
   A smartphone with an app to control the heating system[1] which allows temperature/time settings to be changed.[1] [4]

6. (a) The video feed can be viewed from anywhere in the world.[1] The camera's position can be controlled from anywhere in the world.[1] The video can be analysed (for example for movement)[1] and a notification can be sent to a user.[1] A security light can be turned on / an alarm can be triggered.[1] The video footage can be stored remotely[1] preventing an intruder from destroying it[1] (traditional CCTV systems will often store the footage on a hard drive in the same office). [2]
   (b) If the Internet/network connection fails then the cameras won't work.[1]
   The cameras may be more expensive than more traditional CCTV cameras.[1]
   There may be additional setup / a subscription cost to store the footage.[1]
   There will be security/personal data implications[1] as the data is transmitted over the Internet[1] (which is a public network). Encryption will need to be used.[1] [2]

7. A sensor / air bag deployment could detect the air bag has been deployed.[1] Alternatively, a SOS button in the car could be pressed.[1] This would alert the police/ambulance services to the exact location of the accident[1] allowing them to respond faster.[1] Because the exact location is known, an ambulance/police won't waste time going to an incorrect location.[1]
   The company could provide their own app[1] which allows information to be reported for incidents.[1] In the event of an accident, the phone could report information such as driver/licence plate/location[1] allowing the company to easily contact an insurance company,[1] inform friends and family,[1] send another driver to pick up all the packages.[1] [4]

8. The in car entertainment system could use a microphone to listen to instructions/commands that the driver makes.[1] These could then be used to play music tracks/find locations on navigation software/dial phone numbers[1] which would reduce distractions for the driver.[1]
   The car entertainment system would connect to the microphone and speakers allowing the driver to make hands free phone calls[1] which are safer/more convenient for the driver.[1]
   With an IoE system, the driver would be able to stream music[1] and select any particular track/mood they would like.[1]
   With a virtual assistant, the driver would be able to give commands[1] that control other IoE/smart devices such as home heating/opening a garage door/unlocking front door/turning on lights.[1] They would also be able to carry out other functions such as dictating/sending an email/text message or making a to do list.[1] [4]

9. The following shows examples that could be given for the use of devices, advantages and disadvantages. The question will be marked using the levels based marking criteria at the bottom of the page. [9]

Use of devices:
- A barcode scanner may be used to track packages in warehouses or as they are loaded on/off lorries or delivery vans.
- A smartphone's camera may be used to read barcodes/QR codes/2D barcodes.
- A photo of the package may be taken to prove it has been delivered or show the location it has been delivered to.
- A custom made tracking device may be used which has similar capabilities to a smartphone.
- A smartphone/tablet/specialist device may be used to take a signature on delivery.
- Each time an update is given of the packages location/status this can be uploaded using a 4G/5G connection.
- The user can use a web browser to access the current status/delivery data/current location of the drive/estimated deliver time.
- Updates can automatically be sent to the user as email/text message.

Advantages:
- Using automation for updating customers/delivery drivers is more efficient.
- Customers receive regular updates about their packages.
- Lost packages are easier to locate.
- By having Internet connected devices, real-time information and updates are given.
- Customers are often easily able to give instructions directly to delivery drivers via a form on a webpage.
- Customers can change their delivery date.

Credit will be given for any other suitable response.

Disadvantages:
- Customers need to have access to the Internet.
- The system won't work without the Internet.
- Customers may not have the digital skills/confidence to use the system effectively.
- There is an increased risk of hacking or identity theft.
- If the delivery company suffered a cyber-security breach the impact could be significant.

Credit will be given for any other suitable response.

Level 3 (high) 7-9 marks
A **thorough** discussion which shows **detailed** understanding:
- **Detailed** knowledge and understanding of the use of devices to track products.
- **More than** one advantage and disadvantage of using IoE to track products are **explained**.
- Relevant and appropriate examples are given.
- Consistently used appropriate terminology.

Level (2) (mid) 4-6 marks
An adequate discussion which shows sound understanding:
- **Sound** knowledge and understanding of the use of devices to track products.
- **At least** one advantage **and/or** one disadvantage of using IoE to track products **are described**.
- Some relevant examples are provided although these may not always be appropriate.
- Some use of appropriate terminology.

Level (1) (low): 1-3 marks
A brief discussion which shows limited understanding:
- **Limited** knowledge and understanding of the use of devices to track products.
- **Few** advantage(s) and/or disadvantage(s) of using IoE to track products in general are **identified**.
- Little or no use of appropriate terminology.

0 marks
- No response worthy of credit.

LEVELS-BASED MARK SCHEME FOR EXTENDED RESPONSE QUESTIONS

Example level descriptors

Each exam paper will have an extended response question, such as a discuss question, which is marked by the following levels.

Level	Marks	Level descriptors
HIGH 3	7–9	• A **thorough** discussion with detailed understanding. • A detailed knowledge. • **More than** one advantage **and** one disadvantage are **explained**. • Relevant and appropriate examples are given. • Terminology is consistently used.
MID 2	4–6	• An **adequate discussion** showing sound understanding. • At least **one** advantage and/or **one** disadvantage are **described**. • Some relevant examples are provided although these may not always be appropriate. • Some use of appropriate terminology.
LOW 1	1–3	• A brief discussion showing limited understanding. • Few advantage(s) and/or disadvantage(s) are identified. • Little or no use of appropriate terminology.
0	0	• No response worthy of credit.

Create/draw questions

Each exam paper will have a question that requires a hand-drawn visual solution, such as a mind map, flow chart or visualisation diagram. These questions will be marked with up to 4 marks for layout and up to 4 marks for content. The marks are added giving up to 8 marks.

Marks for layout
- Wholly relevant layout (4 marks)
- Mostly suitable layout (3 marks)
- Simplistic layout (2 marks)
- Minimal layout for the scenario (1 mark)

Marks for content
- All relevant content (4 marks)
- Mostly relevant content (3 marks)
- Some relevance (2 marks)
- Limited relevance (1 mark)

0 marks – no answer worthy of credit

NOTES, DOODLES AND EXAM DATES

Doodles

Key dates

INDEX

Symbols

4G and 5G 51

A

adware 34
Android 18
ANPR (Automatic Number Plate Recognition) 65
anti-malware software 38
audience 53
audio 45

B

backups 37, 38
baiting 35
banking 13, 15
biometric devices 37
black hat 33
bluetooth 52, 56
books 27
Boolean 23
botnet 34
branch 4

C

capacitive touch screens 16
Chrome OS 18
Cloud 49
  storage 28
collaboration tools 44
command line interface 19
communication 44
Computer Misuse Act 40
connectivity 51
Copyright, Designs and Patents act 40
CPU (Central Processing Unit) 17
cybersecurity 33, 36

D

data 23, 57
  destruction 39
  sanitisation 39
  storage 28
  theft 36
  type check 25
databases 20, 46
Data Protection Act (DPA) 41
decision 2
degaussing 39
demographics 53
Denial of Service (DoS) 34, 36
Desktop Publishing (DTP) 47
digital
  communication 44
  devices 48
  platforms 20
displays 16
  size 17
display screen equipment regulations 42
distribution channel 49

E

electronic paper 16
email 26, 49
embedded systems 14, 15
emergency services 63
encryption 38
energy management 60
entertainment 13, 15
ethernet cables 51
extreme test data 30

F

fibre optic cables 51
firewall 37, 38
fitness 14, 15
flash drives 29
flow chart 2
format check 25
Freedom of Information Act (FoIA) 41

G

gestures 21
government statistics 27
Graphical User Interface (GUI) 17, 19
grey hat 33

H

hacker 33
hard disk drive (HDD) 28
hardware 16
health 61
Health and Safety at Work Act 42
home appliances 13, 15
Human Computer Interface (HCI) 12

I

identity theft 36
infographics 44
information 23
input mask 25
intellectual property (IP) 40
Internet 57
  of Everything (IoE) 56
  of Things (IoT) 56
interviews 26
invalid test data 30
iOS 18

K

keyboards 21
keypads 37

L

leaflets 44
legislation 33, 40
length check 25
library mind maps 5
Light Emitting Diodes (LED) 16
Linux 18
liquid crystal displays (LCD) 16
lookup 25

M

macOS 18
magnetic wipe 39
malware 34
manufacturing 62
messaging 49
military 63
mind maps 4
  library mind maps 5
  presentation mind maps 5
  tunnel timeline 6
mobile apps 20, 50
multimedia 50
multi-tasking 18

N

network-attached storage (NAS) 29
network drives 28
newsletters 44
NFC (Near Field Communication) 56
node 4

O

OLED (Organic LED) 16
operating system 12, 18

P

people 57
personal data 41
phishing 35
pillars of IoE 57
Point of Sale (POS) 14
presence check 25
presentation mind maps 5
presentations 44, 46
pretexting 35
primary data collection 26
processes 57

Q

queries 46
questionnaires 26
quid pro quo 35

R

Radio-frequency identification (RFID) 37
RAM (Random Access Memory) 17
range check 25
ransomware 34
reports 44, 46
retail 14, 15
RFID (Radio Frequency Identification) 56

S

scareware 35
secondary data collection 27
security 60
shoulder surfing 35
smart
  boards 48
  devices 56, 64
  phones 13, 48
  TV 48
social
  engineering 35
  media 45
software 46
solid state drive (SSD) 28
spreadsheets 20, 46
spyware 34
storage of data 28

T

tablets 48
technical testing 31
terminator 2
testing 30
things 57
threats 34
touch 21
  screens 16
transport 65
Trojan horse 34
tunnel timeline mind maps 6
two-factor authentication (2FA) 38

U

Ubuntu 18
Unix 18
USB flash drives 29
user
  interaction 21
  testing 31

V

validation 24, 25
valid test data 30
verification 24
video 45
virus 34
visualisation diagrams 7
voice control 21
Voice over Internet Protocol (VoIP) 45, 49

W

websites 20, 27, 50
white hat 33
wi-fi 52
WIMP environment 19
wired connections 51
wireframe 8
wireless connections 51
word processors 46
World Wide Web (WWW) 57
worm 34

Z

zigbee 56

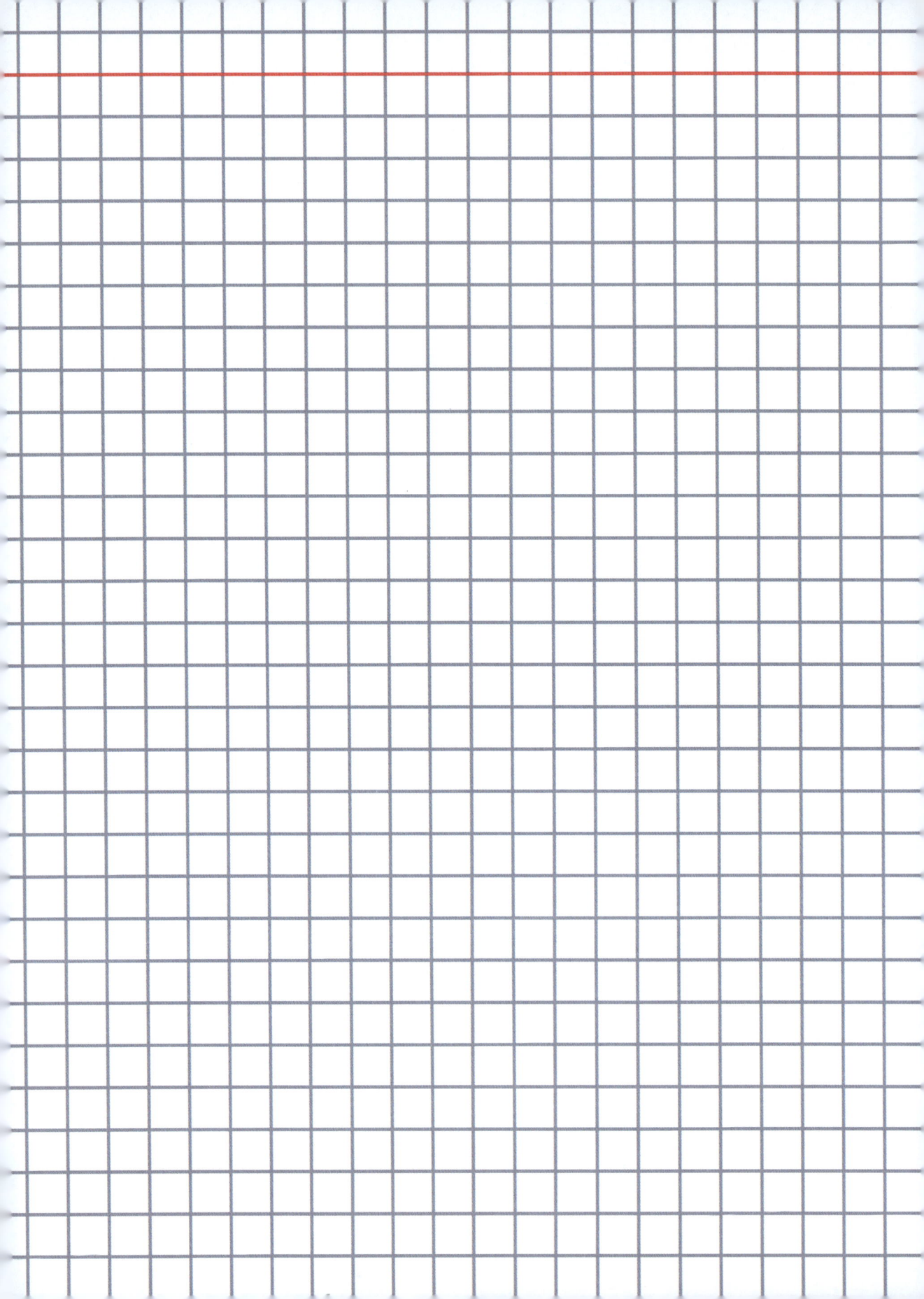

EXAMINATION TIPS

With your examination practice, use a boundary approximation using the following table. Be aware that boundaries are usually a few percentage points either side of this.

Level	Level 2				Level 1		
Grade	Distinction*	Distinction	Merit	Pass	Distinction	Merit	Pass
Code	2*	D2	M2	P2	D1	M1	P1
Boundary	80%	70%	60%	50%	40%	30%	25%

1. Be prepared with a black pen and a ruler.
2. Always read each question carefully. Make sure you understand what the question is asking and follow the instructions. You cannot get marks for giving an answer to a question you think is appearing rather than the actual question.
3. Section B is based around a scenario. Remember to link your answers to this scenario if required.
4. Avoid simply rewriting the question or repeating examples that are already given in the question.
5. It is better to use generic terms such as heart rate monitor or smart watch, rather than brand names such as FitBit.
6. Remember that explain questions have two marks. You need to make a point for the first mark, and then expand this point with a linked development for the second mark. To help you develop your responses, aim to include words such as 'because' or 'therefore'.
7. On describe or explain questions remember to construct your answer in a logical manner.
8. There is one long answer question on the exam paper which is worth 9 marks and could use the command words analyse, discuss, or evaluate. Remember that the answers to these questions need both advantages and disadvantages, and an 'evaluate' question also needs a conclusion.
9. Answer questions in the spaces provided. If this is not possible e.g. due to deleting a wrong answer, indicate the location of the corrected answer on the paper (e.g. 'see next page' or 'my answer is on the last blank page').
10. Cross out any errors neatly.
11. Don't spend too much time on one question or leave any questions unanswered.
12. Make sure your handwriting is clear and legible.
13. Don't let your nerves get the better of you. Remember that you have prepared well, and you can do this.
14. Lastly, try to relax, breathe deeply, and focus on the task at hand. Don't compare yourself to others or worry about what they are doing.

Good luck!

New titles coming soon!

These guides are everything you need to ace your exams and beam with pride. Each topic is laid out in a beautifully illustrated format that is clear, approachable and as concise and simple as possible.

They have been expertly compiled and edited by subject specialists, highly experienced examiners, industry professionals and a good dollop of scientific research into what makes revision most effective. Past examination questions are essential to good preparation, improving understanding and confidence.

- Hundreds of marks worth of examination style questions
- Answers provided for all questions within the books
- Illustrated topics to improve memory and recall
- Specification references for every topic
- Examination tips and techniques
- Free Python solutions pack (CS Only)

Absolute clarity is the aim.

Explore the series and add to your collection at **www.clearrevise.com**

Available from all good book shops

ClearRevise — Illustrated revision and practice
AQA GCSE **Business** 8132

ClearRevise — Illustrated revision and practice
OCR **Creative iMedia** Levels 1/2 J834 (R093, R094)

ClearRevise — Illustrated revision and practice
AQA GCSE **English Language** 8700

ClearRevise — Illustrated revision and practice
Edexcel GCSE **History 1HI0** Weimar and Nazi Germany, 1918–39 Paper 3

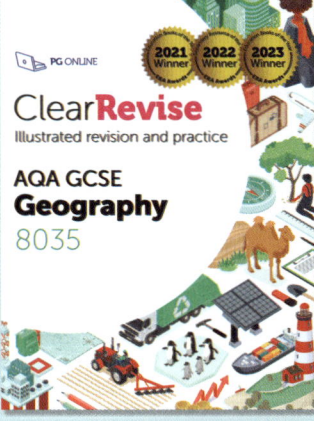

ClearRevise — Illustrated revision and practice
AQA GCSE **Geography** 8035

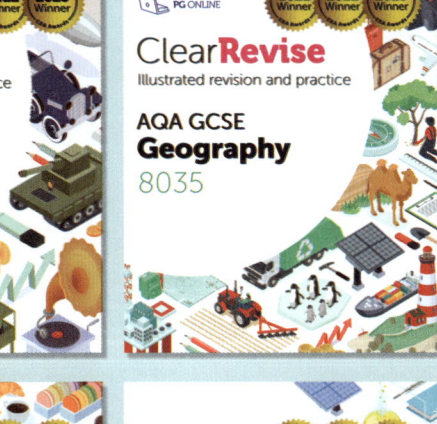

ClearRevise — Illustrated revision and practice
OCR GCSE **Computer Science** J277

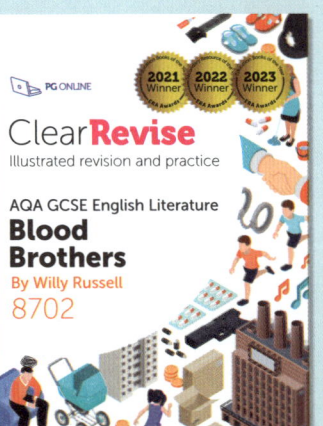

ClearRevise — Illustrated revision and practice
AQA GCSE English Literature **Blood Brothers** By Willy Russell 8702

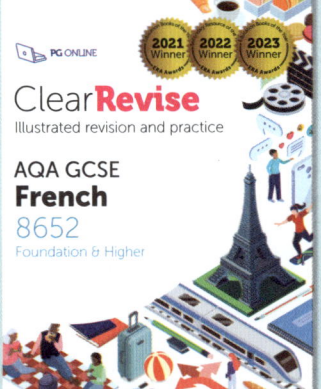

ClearRevise — Illustrated revision and practice
AQA GCSE **French** 8652 Foundation & Higher

ClearRevise — Illustrated revision and practice
AQA GCSE **Combined Science** Trilogy 8464 Foundation & Higher

ClearRevise — Illustrated revision and practice
AQA GCSE **Design and Technology** 8552